Building Blocks of
Quantum Mechanics

Building Blocks of Quantum Mechanics

Theory and Applications

Tao Xiang

CRC Press
Taylor & Francis Group
Boca Raton London New York

CRC Press is an imprint of the
Taylor & Francis Group, an **informa** business

First edition published 2022
by CRC Press
6000 Broken Sound Parkway NW, Suite 300, Boca Raton, FL 33487-2742

and by CRC Press
4 Park Square, Milton Park, Abingdon, Oxon, OX14 4RN

CRC Press is an imprint of Taylor & Francis Group, LLC

ISBN: 978-1-032-00610-9 (hbk)
ISBN: 978-0-367-77150-8 (pbk)
ISBN: 978-1-003-17488-2 (ebk)

DOI: 10.1201/9781003174882

Typeset in Nimbus Roman
by KnowledgeWorks Global Ltd.

Publisher's note: This book has been prepared from camera-ready copy provided by the authors.

Dedication

_____

To my mother.

Contents

Preface

The object of this textbook is to provide a concise yet comprehensive introduction to the principles, concepts, and methods of quantum mechanics. It covers the basic building blocks of quantum mechanics theory and applications, illuminated throughout by accurate explanation, physical insight, and intelligible examples, so that students can easily focus on. After a brief outline of the innovative ideas that lead up to the quantum theory, the book introduces the Schrödinger equation, theory of representation, physical meaning of wave functions, measurement postulate, uncertainty principles, and the standard quantum formalism, like the physical observables and their expectation values, the Schrödinger and Heisenberg pictures, symmetry and conservation laws, electromagnetic interaction and gauge invariance, angular momentum, adiabatic and diabatic evolutions of quantum states, entanglement and identical particles, and relativistic quantum mechanics. The theory is materialized and enriched by including many typical and heuristic examples of quantum mechanics, such as the one-dimensional eigen-problems, harmonic oscillator, Aharonov-Bohm effect, Landau levels, hydrogen atom, Landau-Zener transition and Berry phase. The approximate methods commonly used in quantum mechanics, such as the WKB approximation, variational principle, stationary and time-dependent perturbation theories, are also introduced.

This book results from an undergraduate course given at the University of Chinese Academy of Sciences. With careful development of concepts and thorough explanations, the book is aimed to make quantum mechanics accessible for students majoring in physics, chemistry or electronic engineering and professionals in mathematics and computer sciences. It is self-contained, suitable for a semester course for both junior and senior students. It is assumed that the reader is reasonably familiar with atomic structure, electromagnetism, differential equations, and linear algebra. It may also serve as a beginner's reference book for self-studying Quantum Mechanics.

Notations

If without particular specification, the following conventions are used.

1. μ: mass of a particle
2. Ψ: wave function
3. $\mathbf{r} = (x, y, z)$: spatial coordinate
4. $\mathbf{p} = (p_x, p_y, p_z)$: momentum operator
5. H: Hamiltonian operator
6. ρ: probability density
7. $\mathbf{j} = (j_x, j_y, j_z)$: probability current density or electric current density
8. $\mathbf{J} = (J_x, J_y, J_z)$: angular momentum operator
9. $\mathbf{L} = (L_x, L_y, L_z)$: orbital angular momentum operator
10. $\mathbf{S} = (S_x, S_y, S_z)$: spin operator
11. $\boldsymbol{\sigma} = (\sigma_x, \sigma_y, \sigma_z)$: Pauli matrices
12. m: eigenvalue of the z-component of angular momentum
13. $\mathbf{A}$: vector gauge potential of electromagnetic field
14. φ: Coulomb potential of electromagnetic field
15. $\mathbf{E}$: vector electric field
16. $\mathbf{B}$: vector magnetic field
17. $\mathscr{A}$: Berry connection or fictitious gauge vector
18. $\mathscr{B}$: fictitious magnetic field associated with the Berry phase

Sections marked with $*$ at the end of the section title are for advanced study only.

Formulas in SI units and Gaussian units

SI units are used in this book. Below is a table that shows how the basic formulas presented in this book are converted from SI units to Gaussian units.

	SI unit	Gaussian unit
Schrödinger Hamiltonian	$H = \dfrac{1}{2\mu}(\mathbf{p} - e\mathbf{A})^2 + e\varphi$	$H = \dfrac{1}{2\mu}\left(\mathbf{p} - \dfrac{e}{c}\mathbf{A}\right)^2 + e\varphi$
Dirac Hamiltonian	$H = c\alpha \cdot (\mathbf{p} - e\mathbf{A}) + \alpha_0 \mu c^2 + e\varphi$	$H = c\alpha \cdot \left(\mathbf{p} - \dfrac{e}{c}\mathbf{A}\right) + \alpha_0 \mu c^2 + e\varphi$
Probability current density	$\mathbf{j} = \dfrac{i\hbar}{2\mu}(\Psi\nabla\Psi^* - \Psi^*\nabla\Psi)$ $- \dfrac{e}{\mu}|\Psi|^2\mathbf{A}$	$\mathbf{j} = \dfrac{i\hbar}{2\mu}(\Psi\nabla\Psi^* - \Psi^*\nabla\Psi)$ $- \dfrac{e}{\mu c}|\Psi|^2\mathbf{A}$
Canonical momentum	$\mathbf{p} - e\mathbf{A}$	$\mathbf{p} - \dfrac{e}{c}\mathbf{A}$
Coulomb potential	$V(r) = \dfrac{1}{4\pi\varepsilon_0}\dfrac{q_1 q_2}{r}$	$V(r) = \dfrac{q_1 q_2}{r}$
Electric field	$\mathbf{E} = -\nabla\varphi - \dfrac{\partial \mathbf{A}}{\partial t}$	$\mathbf{E} = -\nabla\varphi - \dfrac{1}{c}\dfrac{\partial \mathbf{A}}{\partial t}$
Magnetic field	$\mathbf{B} = \nabla \times \mathbf{A}$	$\mathbf{B} = \nabla \times \mathbf{A}$
Lorentz force	$\mathbf{F} = q(\mathbf{E} + \mathbf{v} \times \mathbf{B})$	$\mathbf{F} = q\left(\mathbf{E} + \dfrac{1}{c}\mathbf{v} \times \mathbf{B}\right)$
Gauge transformation	$\mathbf{A} \to \mathbf{A} + \nabla\chi$ $\varphi \to \varphi - \dfrac{\partial \chi}{\partial t}$ $\Psi \to \exp\left(\dfrac{ie}{\hbar}\chi\right)\Psi$	$\mathbf{A} \to \mathbf{A} + \nabla\chi$ $\varphi \to \varphi - \dfrac{1}{c}\dfrac{\partial \chi}{\partial t}$ $\Psi \to \exp\left(\dfrac{ie}{\hbar c}\chi\right)\Psi$
Magnetic flux quantum	$\Phi_0 = \dfrac{h}{e}$	$\Phi_0 = \dfrac{ch}{e}$
Fine structure constant	$\alpha = \dfrac{e^2}{4\pi\varepsilon_0\hbar c}$	$\alpha = \dfrac{e^2}{\hbar c}$
Bohr magneton	$\mu_B = \dfrac{e\hbar}{2\mu}$	$\mu_B = \dfrac{e\hbar}{2\mu c}$

Table of fundamental constants

	Symbol	Value
Planck's constant	h	6.62618×10^{-34} J s
Velocity of light	c	2.99792×10^{8} m s^{-1}
Charge of electron	e	1.60219×10^{-19} C
Mass of electron	m_e	9.10953×10^{-31} kg
Ratio of proton to electron mass	m_p/m_e	1836.15
Permittivity of vacuum	ε_0	8.85419×10^{-12} F m^{-1}
Permeability of vacuum	μ_0	1.25665×10^{-6} H m^{-1}
g-factor of the electron	g	2.002319304386
g-factor of the proton	g	5.5858
g-factor of the neutron	g	-3.8261

1 Introduction

1.1 BRIEF HISTORY OF QUANTUM MECHANICS

Quantum mechanics is the fundamental theory of nature at small scales and low energies of atoms and subatomic particles. It, together with Einstein's theory of relativity, is the two major discoveries of the 20th century. They are also the pillars of modern science.

Quantum mechanics is inescapable. Nearly all fields of modern physics, from elementary particles to the big bang, from semiconductors to solar energy cells, are related to quantum physics. It is undoubtedly one of the greatest intellectual achievements of the history of humankind, probably the greatest of those that will remain for centuries.

The phrase "quantum mechanics" was coined (in German, Quantenmechanik) by Max Born, Werner Heisenberg, and Wolfgang Pauli at the University of Gottingen in the early 1920s, and was first used in Born's 1924 paper "Zur Quantenmechanik".

Two experimental observations played central roles in the establishment of quantum mechanics. One is the the black-body radiation first reported by Balfour Stewart in 1858 and then by Gustav Kirchhoff in 1859. It was the solution to this problem by Max Planck in 1900 that led to the birth of quantum theory. Planck is also considered to be the father of the quantum physics. The other is the optical spectrum of hydrogen. The basic theory of quantum mechanics, including the Schrödinger equation, was established based on the solution of this problem.

Major progress in the discovery of quantum mechanics includes:

1. In 1900, Max Planck postulated a quantized energy formula of light to solve the black-body radiation problem. He pointed out that in order to solve the black-body radiation problem, the energy ε of a light oscillator of a given frequency v should not vary continuously, instead it could only take on some discrete values, which are integer multiplies of v

$$\varepsilon = hv, \tag{1.1}$$

where h is a fundamental constant which is called the Planck's constant. Based on this assumption, he found that the spectral density of electromagnetic radiation, emitted by a black body in thermal equilibrium, depends only on the temperature T,

$$I(v) = \frac{8\pi h}{c^3} \frac{v^3}{e^{hv/k_B T} - 1}, \tag{1.2}$$

where k_B is the Boltzmann's constant and c is the speed of light. This formula is also called Planck's law.

Planck's discovery of light quantization was totally at variance with classical physics. It unveiled the mystery of quantum world.

DOI: 10.1201/9781003174882-1

2. In 1905, Albert Einstein adopted Planck's idea and proposed a quantum theory to explain the photoelectric effect discovered by Heinrich Hertz in 1887. He assumed that the electromagnetic field was quantized, called light quanta, with an energy given by Eq. (1.1). He won the 1921 Nobel Prize in Physics for this work. In 1926, the physical chemist Gilbert N. Lewis coined the name photon for these light quanta.

3. In 1913, Niels Bohr developed the Bohr model of the atom, in which he proposed that energy levels of electrons are discrete, hence quantized, and that the electrons revolve in stable orbits around the atomic nucleus but can jump from one energy level (or orbit) to another.

Bohr assumed that an electron in an atom moves in an orbit, which is stationary and not radiative, and its orbital angular momentum is quantized

$$L = \mu v r = n\hbar, \tag{1.3}$$

where μ, v, and r are the mass, the velocity, and the radius of the orbit in which the electron travels. n is an integer and $\hbar = h/2\pi$. For a hydrogen-like atom, the attractive Coulomb force acting on the electron by the nucleus of charge Ze, is equated with the centripetal acceleration (v^2/r), giving by

$$\frac{\mu v^2}{r} = \frac{Ze^2}{4\pi\varepsilon_0 r^2}. \tag{1.4}$$

From these two equations, v and r are then found to be

$$v = \frac{Ze^2}{4\pi\varepsilon_0 \hbar n}, \qquad r = \frac{4\pi\varepsilon_0 \hbar^2 n^2}{Z\mu e^2}. \tag{1.5}$$

The total energy of the electron, which is a sum of the kinetic energy and the potential energy, is now quantized

$$E_n = \frac{1}{2}\mu v^2 - \frac{Ze^2}{4\pi\varepsilon_0 r} = -\frac{\mu}{2\hbar^2}\left(\frac{e^2}{4\pi\varepsilon_0}\right)^2 \frac{1}{n^2}. \tag{1.6}$$

This formula offers an excellent explanation to the radiation spectral lines of hydrogen. This theory earned Bohr the 1922 Nobel Prize. Its two most important features have even survived in the present-day quantum mechanics: (1) the existence of stationary states and (2) the radiation frequency equals the energy difference between the initial and final states in a transition.

The first electrical measurement to clearly show the quantum nature of atoms was carried out by Franck and Hertz. They used a beam of mono-energetic electrons to bombard atoms and measured the kinetic energy of the scattered electrons. As long as the kinetic energy of incident electron is below the difference $E_n - E_1$, atoms cannot absorb the energy and all collisions are elastic. As soon as the kinetic energy is larger than this difference, inelastic collisions occur and some atoms go into their first excited states.

Similarly, atoms can be excited into the second excited state as soon as the kinetic energy of the incident electron is larger than $E_3 - E_1$. This was exactly what was found experimentally. Franck and Hertz were awarded the Nobel Prize in Physics for this work.

4. In 1923, the French physicist Louis de Broglie published his theory of wave-particle duality with the claim that particles can exhibit wave characteristics and vice versa. He proposed that matter has wave as well as particle properties, and particularly particles can behave as waves and their wavelength λ is related to the momentum p of the particle by

$$\boxed{\lambda = \frac{h}{p}.} \tag{1.7}$$

λ is called the de Broglie wavelength.

De Broglie's formula was confirmed three years later for electrons (which differ from photons in having a rest mass) with the observation of electron diffraction in two independent experiments. At the University of Aberdeen, George Paget Thomson passed a beam of electrons through a thin metal film and observed the predicted interference patterns. At Bell Labs, Clinton Joseph Davisson and Lester Halbert Germer guided their beam through a crystalline grid.

5. In 1925, Heisenberg wrote his first paper on quantum mechanics in the matrix representation. Quantum mechanics was born when the German physicists Werner Heisenberg, Max Born, and Pascual Jordan developed matrix mechanics that year.

6. In 1926, the Austrian physicist Erwin Schrödinger invented wave mechanics and the non-relativistic Schrödinger equation. He did this by transforming the vague idea of de Broglie into a theory of wave mechanics. He did this work in the Christmas season of 1925 and published it in 1926. By solving this equation with a potential given by the Coulomb potential of a hydrogen atom, he found that only certain discrete values of energy lead to acceptable wave functions. The values of energy depend only on the integers that are identical with those given by the Bohr theory.

   In 1926, Schrödinger showed that, in spite of apparent dissimilarities of his wave mechanics with Heisenberg's matrix mechanics, the two theories are equivalent mathematically.

7. In 1926, Max Born proposed his probability interpretation of the wave function. He pointed out that the modulus square of the wave function represents the probability of particle in a particular position or state. This interpretation is a pillar of the Copenhagen interpretation of quantum mechanics.

8. In 1927, Heisenberg formulated the uncertainty principle. The Copenhagen interpretation of this principle started to take shape at about the same time. The uncertainty principle was not accepted by everyone at beginning. Its most outspoken opponent was Einstein.

9. In 1928, British physicist Paul Dirac discovered how quantum theory could be expressed in a form which was invariant under the Lorentz transformation of special relativity. He proposed the Dirac equation which unified quantum theory with special relativity. Based on the equation, he discovered the origin of electron spin and predicted the existence of positron, which is an antiparticle of electron. He also introduced the influential bracket notation and pioneered the use of operator theory, as described in his famous 1930 textbook.

1.2 SCHRÖDINGER EQUATION

Particles have characteristic trait of both particles and waves. This radical principle proposed by de Broglie cannot be reconciled in classical mechanics, since the motions of particles are governed by Newton's law, while those of waves are governed by the wave equation.

The wave equation is a second-order linear hyperbolic partial differential equation

$$\frac{\partial^2 \Psi}{\partial t^2} = v^2 \nabla^2 \Psi, \tag{1.8}$$

where v is the propagation velocity of the wave, Ψ quantifies the amplitude of the wave. This equation provides a good description for a wide range of phenomena and plays an important role in electromagnetism, optics, gravitational physics, and heat transfer.

A solution of the wave equation is the plane wave

$$\Psi(r,t) = e^{-i(\omega t - \mathbf{k} \cdot \mathbf{r})}, \qquad \omega^2 = v^2 \mathbf{k}^2 \tag{1.9}$$

where $\mathbf{r} = (x,y,z)$. ω and $\mathbf{k}$ are the frequency and wave vector of the oscillation, respectively. The wave equation is linear. The sum of any two solutions is again a solution. This property is called the superposition principle.

Inspired by de Broglie's idea of wave-particle duality, it was believed that a wave equation would describe the motion of particle. This was a basis hypothesis Schrödinger made in uncovering the equation named after him. In general, physical solutions may not be purely described by plane waves. So the superposition principle is required. This principle should also be obeyed by a quantum particle if it is described by a wave equation. This implies that the equation governing the motion of a quantum particle must be a linear differential equation.

Schrödinger equation was a magic invention of Schrödinger. This equation was derived from nowhere, except from Schrödinger's mind. It is justified by an exhaustive comparisons of the predictions made based on this equation with experiments and by the successful applications to many problems.

The Schrödinger equation is the basic equation of quantum mechanics. It was not derived from any existing fundamental laws and should simply be taken as a basic postulate of quantum mechanics.

Postulate 1:

A state of a quantum mechanical system is completely specified by a complex wave function $\Psi(\mathbf{r},t)$ that evolves in time according to the time-dependent Schrödinger equation

$$i\hbar\frac{\partial}{\partial t}\Psi(\mathbf{r},t) = \left[-\frac{\hbar^2}{2\mu}\nabla^2 + V(\mathbf{r})\right]\Psi(\mathbf{r},t), \qquad (1.10)$$

where $V(\mathbf{r})$ is the potential energy.

It should be emphasized that the Schrödinger equation is valid only in the non-relativistic limit. More generally, it should be replaced by the Dirac equation for an electron or other spin-1/2 particle, the Klein-Gordon equation for a spin-0 particle, or other equations that satisfies the relativistic covariance.

The Schrödinger equation was established based on the quantization of energy conservation using quantum operators and the de Broglie relation. Below we give some arguments that might be helpful in understanding the meaning of each term in this equation.

To construct an equation governing the motion of quantum particles, it clearly requires that its solution must satisfy the energy conservation and its energy is given by the sum of kinetic and potential energies

$$E = \frac{\mathbf{p}^2}{2\mu} + V(\mathbf{r}), \qquad (1.11)$$

where E is the energy and $\mathbf{p}$ is the corresponding momentum. Moreover, in the free space with $V(\mathbf{r}) = 0$, it is expected that this equation should have a plane-wave-like solution as implied by the wave-particle duality, namely

$$\Psi = Ae^{i(\mathbf{k}\cdot\mathbf{r}-\omega t)} = Ae^{i(\mathbf{p}\cdot\mathbf{r}-Et)/\hbar}. \qquad (1.12)$$

Here $E = \hbar\omega$ and the wave vector $\mathbf{k} = \mathbf{p}/\hbar$ is related to the wave length by the formula

$$|\mathbf{k}| = \frac{2\pi}{\lambda}. \qquad (1.13)$$

The spatial and time derivations of Ψ are given by

$$-i\hbar\nabla\Psi = \mathbf{p}\Psi, \qquad (1.14)$$

$$i\hbar\frac{\partial}{\partial t}\Psi = E\Psi. \qquad (1.15)$$

These equations suggest that E is the energy eigenvalue of the Hamiltonian operator

$$H = i\hbar\frac{\partial}{\partial t}, \qquad (1.16)$$

and $\mathbf{p}$ is the eigenvalue of the momentum operator

$$\mathbf{p} = -i\hbar\nabla = \left(-i\hbar\frac{\partial}{\partial x}, -i\hbar\frac{\partial}{\partial y}, -i\hbar\frac{\partial}{\partial z} \right). \qquad (1.17)$$

Here, for simplicity, we have used the same symbol $\mathbf{p}$ to denote the momentum operator and its eigenvalue. However, physically these two quantities are different.

In Eq. (1.11), if we substitute E and $\mathbf{p}$ by their corresponding operators, we then have

$$H = -\frac{\hbar^2}{2\mu}\nabla^2 + V(\mathbf{r}). \qquad (1.18)$$

The Schrödinger equation is simply the eigen-equation of the Hamiltonian

$$H\Psi = i\hbar\frac{\partial}{\partial t}\Psi. \qquad (1.19)$$

The Schröodinger equation plays the role of Newton's laws and conservation of energy in classical mechanics. It is a linear partial differential equation in terms of the wave function which predicts precisely the probability of events or outcome. If $\Psi(\mathbf{r},0)$ is known at some initial time $t = 0$, the wave function, $\Psi(\mathbf{r},t)$, at any further time t is determined by the equation.

Quantum mechanics differs from classical physics in the following aspects:

1. Energy, momentum and other quantities are often restricted to discrete values, which are referred as quantization.
2. Objects have wave-particle duality.
3. There are limits to the precision with which quantities can be spontaneously determined (uncertainty principle).

The Schrödinger wave formulation of quantum mechanics (also called wave mechanics) is mathematically equivalent to the matrix formulation of quantum mechanics (also called matrix mechanics) created by Werner Heisenberg, Max Born, and Pascual Jordan. This equivalence was first revealed by Schrödinger.

It was believed that classical mechanics could be derived from quantum mechanics as an approximation valid only at large or macroscopic scales. This is so-called Bohr's correspondence principle. This principle was formulated by Niels Bohr in 1920. It demands that classical physics and quantum physics give the same answer when the systems become large.

1.3 PROBABILITY INTERPRETATION OF WAVE FUNCTION

Wave function plays an important role in quantum mechanics. However, it was not clear how to interpret it. Schrödinger originally thought that wave function represents a particle that is spread out in space, with most of its density being where the wave function is large. This was incompatible with the fact that an electron is concentrated in space and never breaks up. He also tried to interpret it as a charge density, but not successful because the wave function could be a complex number.

In 1926, Max Born proposed a probability interpretation of wave function, which relates predictions of quantum mechanics to probabilistic observations:

Postulate 2:

The probability density of finding a particle at a given location and time is proportional to the square of the modulus of the wave function associated with the particle,

$$\rho(\mathbf{r},t) = |\Psi(\mathbf{r},t)|^2 = \Psi^*(\mathbf{r},t)\Psi(\mathbf{r},t). \qquad (1.20)$$

The probability interpretation assumes that the norm of the wave function measures the probability of the system in a particular coordinate or representation, in which the total probability is normalized. Hence, $|\Psi(\mathbf{r},t)|^2 d\mathbf{r}$ is the probability of finding the particle at coordinate r and time t in the volume element $d\mathbf{r}$. This is a fundamental assumption made by Born and part of the Copenhagen interpretation of quantum mechanics. It is also called "Born's rule".

The probability interpretation is an essential ingredient of the Copenhagen interpretation of quantum mechanics. Since the total probability of finding the particle must be unity, the wave function should be normalized so that its spatial integration equals unity

$$\int d\mathbf{r}\,|\Psi(\mathbf{r},t)|^2 = 1, \qquad (1.21)$$

where the integral extends over all space.

A wave function whose probability integration is finite is said to be square integrable and can be always normalized to unity by multiplying an appropriate constant.

Two wave functions, Ψ_1 and Ψ_2, which differ from each other just by a complex multiplicative factor of modulus one, i.e. $\Psi_2 = \exp(i\theta)\Psi_1$, are equivalent, where θ is a coordinate-independent real number. They describe the same quantum state with the same probability density, $|\Psi_1|^2 = |\Psi_2|^2$, and satisfy the same normalization condition, i.e.

$$\int d\mathbf{r}\,|\Psi_1|^2 = \int d\mathbf{r}\,|\Psi_2|^2. \qquad (1.22)$$

The Born's probability rule has been widely accepted in quantum mechanics. However, debates on this probability interpretation have never been ended ever since it was proposed. Albert Einstein, one of the founders of quantum theory, for example, did not accept this interpretation because he believed that it leads to the rejection of determinism and of causality. His famous quotation, "God does not play with dice", was his response to this probability interpretation. Schrödinger also did not like the statistical or probability interpretation of quantum mechanics.

1.4 STATIONARY SCHRÖDINGER EQUATION

The time-dependent Schrödinger equation is greatly simplified if the Hamiltonian H is time independent. It admits the time-dependent wave function $\Psi(r,t)$ to have a simple factorized form

$$\Psi(\mathbf{r},t) = e^{-iEt/\hbar}\Psi(\mathbf{r}). \tag{1.23}$$

Substituting it into Eq. (1.10), we then obtain the time-independent Schrödinger equation

$$\left[-\frac{\hbar^2}{2\mu}\nabla^2 + V(r)\right]\Psi(\mathbf{r}) = E\Psi(\mathbf{r}). \tag{1.24}$$

This is the eigen-equation of the Hamiltonian, E is the eigenenergy and $\Psi(\mathbf{r})$ is the corresponding eigenfunction. Apparently, $\Psi(\mathbf{r},t)$ is an eigenfunction of operator $i\hbar\partial/\partial t$

$$i\hbar\frac{\partial}{\partial t}\Psi(\mathbf{r},t) = E\Psi(\mathbf{r},t). \tag{1.25}$$

1.5 CONSERVATION OF PROBABILITY

From the definition of the probability density, we can find its first order time derivative

$$
\begin{aligned}
\frac{\partial}{\partial t}\rho(\mathbf{r},t) &= \frac{\partial\Psi^*(\mathbf{r},t)}{\partial t}\Psi(\mathbf{r},t) + \Psi^*(\mathbf{r},t)\frac{\partial\Psi(\mathbf{r},t)}{\partial t} \\
&= \frac{\hbar}{2i\mu}\left[\Psi(\mathbf{r},t)\nabla^2\Psi^*(\mathbf{r},t) - \Psi^*(\mathbf{r},t)\nabla^2\Psi(\mathbf{r},t)\right] \\
&= -\frac{i\hbar}{2\mu}\nabla\cdot\left[\Psi(\mathbf{r},t)\nabla\Psi^*(\mathbf{r},t) - \Psi^*(\mathbf{r},t)\nabla\Psi(\mathbf{r},t)\right]. \tag{1.26}
\end{aligned}
$$

This equation can be also expressed as[1]

$$\frac{\partial}{\partial t}\rho(\mathbf{r},t) + \nabla\cdot\mathbf{j}(\mathbf{r},t) = 0, \tag{1.27}$$

where

$$\mathbf{j}(\mathbf{r},t) = \frac{i\hbar}{2\mu}\left[\Psi(\mathbf{r},t)\nabla\Psi^*(\mathbf{r},t) - \Psi^*(\mathbf{r},t)\nabla\Psi(\mathbf{r},t)\right] \tag{1.28}$$

---

[1]The probability conservation law (1.27) is valid in a system where the mass μ is coordinate independent. If μ depends on the coordinate $\mathbf{r}$, the derivative of the mass will generate an extra term in Eq. (1.27)

$$\partial_t\rho(\mathbf{r},t) + \nabla\cdot\mathbf{j}(\mathbf{r},t) = \frac{i\hbar}{2}\left[\Psi(\mathbf{r},t)\nabla\Psi^*(\mathbf{r},t) - \Psi^*(\mathbf{r},t)\nabla\Psi(\mathbf{r},t)\right]\cdot\nabla\frac{1}{\mu(\mathbf{r})}$$

which would break the probability conservation law.

is called the probability current (sometimes it is also called probability flux). It describes the flow of probability in terms of probability per unit time.

Eq. (1.27) is called the equation of probability conservation. Using the momentum operator, the probability current can be also expressed as

$$j(r,t) = \frac{1}{2\mu}[\Psi^*(r,t)\,p\Psi(r,t) - \Psi(r,t)\,p\Psi^*(r,t)].\qquad(1.29)$$

The probability density and the probability current have the following properties:

1. For a complex wave function $\Psi(r,t)$, it can be expressed using its amplitude $R(r,t)$ and phase $\theta(r,t)$ as a complex exponential form

$$\Psi(r,t) = R(r,t)\,e^{i\theta(r,t)},\qquad(1.30)$$

where $R(r,t)$ and $\theta(r,t)$ are real functions of r and t. The square of $R(r,t)$ is just the probability density

$$\rho(r,t) = \Psi^*(r,t)\Psi(r,t) = R^2(r,t).\qquad(1.31)$$

The corresponding probability current is

$$\begin{aligned}j(r,t) &= \frac{i\hbar}{2\mu}R^2(r,t)\left[e^{i\theta(r,t)}\nabla e^{-i\theta(r,t)} - e^{-i\theta(r,t)}\nabla e^{i\theta(r,t)}\right]\\ &= \frac{\hbar}{\mu}\rho(r,t)\nabla\theta(r,t).\qquad(1.32)\end{aligned}$$

This equation indicates that if the wave function is real (up to an arbitrary complex multiplication factor), the probability current is zero. Thus a real function has always zero probability current.

2. If we use the familiar formula of the current operator

$$j(r,t) = \rho(r,t)\,v(r,t),\qquad(1.33)$$

where $v(r,t)$ is the classical velocity of a particle (also the group velocity of the wave), then

$$v(r,t) = \frac{\hbar}{\mu}\nabla\theta(r,t).\qquad(1.34)$$

In other words, the spatial gradient of the phase is proportional to the momentum

$$p(r,t) = \mu v(r,t) = \hbar\nabla\theta(r,t).\qquad(1.35)$$

3. The time derivative of the total probability

$$\frac{d}{dt}\int dr|\Psi(r,t)|^2 = -\int dr\nabla\cdot j(r,t) = -\oint_\infty dS\cdot j(r,t) = 0\qquad(1.36)$$

because there is not any current flowing in or out the system at infinity. This is a consequence of probability conservation.

4. If $\Psi(\mathbf{r},t)$ is the wave function of a particular eigenenergy state, then the probability $|\Psi(\mathbf{r},t)|^2$ does not depend on time t. In this case, the time derivative of the total probability in an arbitrary volume V vanishes and

$$\frac{d}{dt}\int_V d\mathbf{r}|\Psi(\mathbf{r},t)|^2 = -\int_V d\mathbf{r}\nabla\cdot\mathbf{j}(\mathbf{r},t) = -\oint_\Omega d\mathbf{S}\cdot\mathbf{j}(\mathbf{r},t) = 0, \qquad (1.37)$$

where Ω denotes the surface of V. Thus the net probability current flowing in and flowing out an arbitrary volume should be zero.

These properties of the probability current hold in arbitrary dimensions, although the discussion is made based on the formulas derived in three dimensions.

In one dimension, ∇ is replaced by ∂_x. The probability current operator becomes

$$j(x,t) = \frac{i\hbar}{2\mu}\left[\Psi(x,t)\,\partial_x\Psi^*(x,t) - \Psi^*(x,t)\,\partial_x\Psi(r,t)\right]. \qquad (1.38)$$

The corresponding equation of probability conservation is

$$\partial_t\rho(x,t) + \partial_x j(x,t) = 0. \qquad (1.39)$$

1.6 QUANTUM SUPERPOSITION

The Schrödinger equation is a linear partial differential equation. This means that any linear combination of solutions is also a solution of this equation: any two or more quantum states can be superposed to become another quantum state, and conversely, any quantum state can be represented as a sum of two or more other quantum states. This is a fundamental property of the Schrödinger equation. It is referred as quantum superposition.

It is simple to verify the superposition principle. Supposing $\Psi_\alpha(\mathbf{r},t)$ $(\alpha = 1,\cdots n)$ is a solution of the Schrödinger equation

$$i\hbar\frac{\partial}{\partial t}\Psi_\alpha(\mathbf{r},t) = H\Psi_\alpha(\mathbf{r},t), \qquad (1.40)$$

and Ψ is a linear superposition of these wave functions

$$\Psi(\mathbf{r},t) = \sum_{\alpha=1}^{n} C_\alpha\Psi_\alpha(\mathbf{r},t), \qquad (1.41)$$

where C_α is an arbitrary complex number. It is straightforward to show that Ψ is also a solution of the time-dependent Schrödinger equation

$$i\hbar\frac{\partial}{\partial t}\Psi(\mathbf{r},t) = H\Psi(\mathbf{r},t). \qquad (1.42)$$

On the other hand, if $\Psi_\alpha(\mathbf{r})$ $(\alpha = 1,\cdots n)$ is a static eigenfunction of the Hamiltonian with E_α the corresponding eigenenergy,

$$H\Psi_\alpha(\mathbf{r}) = E_\alpha\Psi_\alpha(\mathbf{r}) \qquad (1.43)$$

and

$$\Psi(\mathbf{r}) = \sum_{\alpha=1}^{n} C_\alpha \Psi_\alpha(\mathbf{r}) \tag{1.44}$$

is a superposition of these states, then the evolution of $\Psi(\mathbf{r})$ with time is just given by

$$\Psi(\mathbf{r},t) = \sum_{\alpha=1}^{n} C_\alpha e^{-iE_\alpha t/\hbar} \Psi_\alpha(\mathbf{r}). \tag{1.45}$$

This can be readily verified

$$i\hbar \frac{\partial}{\partial t} \Psi(\mathbf{r},t) = \sum_{\alpha=1}^{n} C_\alpha e^{-iE_\alpha t/\hbar} E_\alpha \Psi_\alpha(\mathbf{r}) = H\Psi(\mathbf{r},t). \tag{1.46}$$

Quantum superposition is one of the most fundamental principles of quantum mechanics. Dirac even believed that one could proceed to build up the theory of quantum mechanics on the basis of quantum superposition with minimum number of other assumptions.[2]

1.6.1 NO CLONING THEOREM

A simple but important consequence of the superposition principle is that it is impossible to create an independent and identical copy of an arbitrary unknown quantum state. This is just the statement of the no cloning theorem.[3] It underlies the security of information encoding of quantum states as well as quantum cryptography.

To see why the superposition principle prohibits the arbitrary cloning of quantum states, let us consider how a quantum state, of which we have no prior knowledge on its detailed structure, is copied. Given a quantum state, Ψ, defined in subspace A, to clone is to generate an exact copy of this state in another subspace of the same size, say subspace B. For doing this, we prepare a simple state ϕ_0 in B so that the whole system is initially in a tensor product state $\Psi \otimes \phi_0$. After cloning (assuming we know how to do it), ϕ_0 becomes Ψ and the state of the whole system becomes $\Psi \otimes \Psi$. Hence if U is the quantum operator that performs cloning for all quantum states, then

$$U(\Psi \otimes \phi_0) = \Psi \otimes \Psi. \tag{1.47}$$

Let Ψ_a and Ψ_b be two arbitrary states in A, to clone them means

$$U(\Psi_a \otimes \phi_0) = \Psi_a \otimes \Psi_a, \tag{1.48}$$
$$U(\Psi_b \otimes \phi_0) = \Psi_b \otimes \Psi_b. \tag{1.49}$$

Now consider a linear superposition of these two states $\Psi_c = \Psi_a + \Psi_b$. The linearity of the superposition principle implies that the copy of Ψ_c is simply a sum of the copies of Ψ_a and Ψ_b, hence

$$U(\Psi_c \otimes \phi_0) = U(\Psi_a \otimes \phi_0) + U(\Psi_b \otimes \phi_0) = \Psi_a \otimes \Psi_a + \Psi_b \otimes \Psi_b. \tag{1.50}$$

---

[2] P. Dirac, The Principles of Quantum Mechanics, Oxford University Press, 1930.

[3] W. K. Wootters and W. H. Zurek, Nature **299**, 802 (1982); D. Dieks, Phys. Lett. A **92**, 271 (1982).

On the other hand, if Ψ_c is directly cloned, we should have

$$U(\Psi_c \otimes \phi_0) = \Psi_c \otimes \Psi_c = \Psi_a \otimes \Psi_a + \Psi_a \otimes \Psi_b + \Psi_b \otimes \Psi_a + \Psi_b \otimes \Psi_b, \quad (1.51)$$

which is clearly not equal to $\Psi_a \otimes \Psi_a + \Psi_b \otimes \Psi_b$. Thus there is no reliable way to clone all quantum states. The reason for this is because the cloning operator U is not linear. Of course, the no-cloning theorem does not preclude the construction of any known quantum state.

1.6.2   SCHRÖDINGER CAT

Superposition principle exists in both quantum and classical wave mechanics. However, the physical consequence of quantum superposition is much more significant than the classical one. It is mysterious why ordinary macroscopic objects do not display this quantum feature of superposition. In 1935, Schrödinger devised a well-known thought experiment, now known as Schrödinger's cat paradox, to question this principle. It presents a "classical" cat that evolves over time into a superposition of classically distinct quantum states (alive and dead), as a result of interaction with a random subatomic event that may or may not occur.[4] This interaction between the system (cat here) and the environment (radioactive atom) is called "entanglement". Hence the cat and the radioactive atom are "entangled".

1.7   OPERATORS

Every dynamical variable of observable, that is measurable in experiments, is associated with a linear operator. This is a basic assumption of quantum mechanics.

Postulate 3:
To every observable in classical mechanics, there corresponds a linear, Hermitian operator in quantum mechanics.

For example, energy is associated with the Haimiltonian and momentum is associated with the momentum operator. This assumption can be also stated as any observable dynamic quantity that can be measured by experiments, like momentum or position of a particle, can be represented as a linear hermitian[5] operator.

An operator O is linear if it has the property

$$O(c_1 \Psi_1 + c_2 \Psi_2) = c_1(O\Psi_1) + c_2(O\Psi_2), \quad (1.52)$$

---

[4]The Schrödinger cat provides a cartoon picture for illustrating quantum superposition. However, it should be pointed out that the Schrödinger cat is not truly a paradox, since the cat is not really in a superposition state (both alive and dead) if it was alive before the box is closed. Before the radioactive atom decays, the cat is always alive. After the decay, the cat dies, again not in a superposition state. The only thing we do not know is the time the radioactive atom decay. Not knowing whether the cat is alive or dead does not mean that the cat is in a superposition state if not considering the role of radioactive atom.

[5]Properties of Hermitian operators are discussed in 3.1.3

where Ψ_1 and Ψ_2 are two wave functions, c_1 and c_2 are complex constants.

Many mathematical or physical operations (or transformations) are not linear. For example, the square is not a linear operator

$$O(c_1\Psi_1 + c_2\Psi_2) = (c_1\Psi_1 + c_2\Psi_2)^2 \neq c_1(O\Psi_1) + c_2(O\Psi_2). \qquad (1.53)$$

If operator O is a function of the coordinate and momentum operators, then in the configuration space it can be represented as

$$O(\mathbf{r},\mathbf{p}) = O(\mathbf{r}, -i\hbar\nabla). \qquad (1.54)$$

It is obtained by performing the substitution $\mathbf{p} \rightarrow -i\hbar\nabla$. This holds if more than one particle is present.

1.8 QUANTUM MEASUREMENT

The eigenvectors, also call eigenstate, of a physical observable form an orthonormal basis space. For a particular physical observable, in order to know its value in a quantum state, a measurement has to be taken. In a classical system, one can measure the position, velocity, or other physical quantities of a particle without disturbing its state of motion. In a quantum world, as the state could be in an arbitrary superposition of the eigenstates of the operator that is measured. This implies that the outcome of measurements is probabilistic. A basic postulate of measurement is

> Postulate 4:
> In any measurement of the observable associated with an operator, the only values that will ever be observed are the eigenvalues of this operator. At each individual measurement, only one of the eigenvalues of the operator will be observed. If the i'th eigenvalue is observed, the system will completely collapse onto the corresponding eigenstate immediately after the measurement.

This postulate is also referred to as the Copenhagen interpretation of quantum measurement. It implies that individual measurements of a quantum system have multiple outcomes. It is unknown which of these eigenvalues will be observed a priori. We only know in a probabilistic way the relative opportunities to realize any particular eigenvalue. The probabilities become meaningful only in the limit of large numbers of measurements.

Moreover, any measurement performed on a quantum system necessarily involves some interaction between the system and the observer. This will inevitably bring about a change of the quantum state. An important consequence of a measurement is that it induces a transition from a probabilistic to a deterministic outcome. A subsequent measurement immediately after the first one for the same physical observable will yield the same eigenvalue but with a probability one. This is known as the "reduction of the wave packet" or the "collapse of the wave function". Due to the "collapse", it is clearly that the energy is generally not conserved in a quantum measurement.

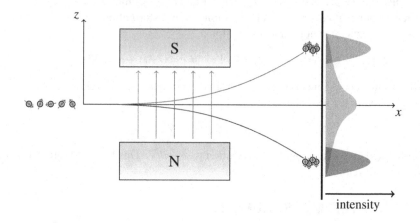

Figure 1.1 Schematic plot of the Stern-Gerlach experimental apparatus (left) and the measurement intensity (right).

For the Schrödinger cat which is assumed to be in a superposition state of "alive" or "dead", once the box is open, the cat is "observed" either alive (if no radioactive decay) or dead (if radioactive decay happened). This observation realizes a measurement that induces the collapse of the Schrödinger cat, which is in a superposition state of alive and dead, onto one of the "eigenstate" (here alive or dead) of the "observer" (human eyes). The collapse is a consequence of the interplay between a microscopic quantum world and a macroscopic observer.

1.8.1   STERN-GERLACH EXPERIMENT

The Stern-Gerlach experiment, first conducted by Stern and Walther Gerlach at the University of Frankfurt in 1922, has played an important role in illuminating the measurement postulate. In the experiment, as schematically illustrated in Fig. 1.1, silver atoms were sent from an oven along the x-axis through a spatially varying magnetic field. The atoms were deflected before arriving at a detector screen. The deflection results from the force applied by the magnetic field gradient to the magnetic moment of a Silver atom.

For simplicity in the discussion below, we assume that all the atoms leaving the oven have the same speed and move along the x-axis direction, and both the magnetic field and its gradient point exactly along the z-axis direction. A silver atom possesses a finite magnetic dipole moment

$$\mathbf{m} = \frac{e}{2\mu}\mathbf{L}, \tag{1.55}$$

where $\mathbf{L}$ is the angular momentum. Its interaction with the applied magnetic field yield two effects:

1. The magnetic field imposes a toque on the magnetic dipole, so that the dipole will precess about the direction of the magnetic field. The x and y-components of **m** will change with time, but this will not affect the motion of the z-component.

2. More importantly, a nonuniform magnetic will exert a force on the atom. The force applying to the dipole by an external magnetic field, $\mathbf{B} = B(z)\hat{z}$ pointing along the z-axis, is given by

$$f = -\frac{dU}{dz} = m_z \frac{dB(z)}{dz} \tag{1.56}$$

where m_z is the z-component of the magnetic dipole moment and

$$U = -\mathbf{m} \cdot \mathbf{B} = -m_z B(z) \tag{1.57}$$

is the potential energy of the Silver atom in the magnetic field. The force is pointing along the z-axis, determined by the gradient of the magnetic field.

In the thermal environment of the oven, the initial dipole moments are randomly oriented. In the classical picture, it is expected that the deflected Silver atoms should show a continuous distribution, whose intensity is qualitatively represented by the gray area in Fig. 1.1, on the detector screen. The deflection angle increases with the absolute value of m_z.

Surprisingly, the observed distribution of deflected Silver atoms is separated into two parts, schematically represented by the light red and blue peaks in Fig. 1.1, respectively. It suggests that only two of the possible values of m_z can be detected no matter how m_z is initially prepared. In other words, each of two allowed values of m_z corresponds to a quantum state, and each atom is in a superposition state of these two quantum state. Assuming Φ_1 and Ψ_2 are these two allowed states, then the state initially prepared is

$$\Psi = c_1 \Psi_1 + c_2 \Psi_2. \tag{1.58}$$

After measurement, Ψ collapses either onto Ψ_1 or onto Ψ_2, losing the memory of its initial state. But the probabilities to find the atom in the first and second states are given by $|c_1|^2$ and $|c_2|^2$, respectively (assuming Ψ_1, Ψ_2 and Ψ are all normalized).

The reason to use Silver atoms to do the experiment is because Silver atoms have finite magnetic moments, but are electrically neutral. This avoids the deflection induced by the Lorentz force applying to a charged particle moving through a magnetic field.

The experiment provided the first evidence that electron has an intrinsic angular momentum — spin. However, the original experiment had nothing to do with the discovery of electronic spin. The spin of electron was first discovered by Uhlenbeck and Goudsmit in 1925, based on the analysis of atomic spectra. What the experiment was intended to test was a "space-quantization" associated with the orbital angular momentum of electrons, predicted by the old quantum theory of Bohr and Sommerfeld. Later, it was realized that the original interpretation of the experiment was wrong, and it was the spin of electron that was observed.

The most important information revealed by the Stern-Gerlach experiment is that only one component of the spin is measured at one time, in agreement with the Copenhagen interpretation of quantum measurement.

1.9 EXPECTATION VALUES

In quantum mechanics, an expectation value, also referred to as a mean value, is a probabilistic expected value of a physical observable. It is an average of all the possible outcomes of experimental measurements, weighted by their probabilities, and a fundamental concept of quantum mechanics.

Let $\Psi(\mathbf{r})$ be a normalized wave function. As $\rho(\mathbf{r})d\mathbf{r} = \Psi^*(\mathbf{r})\Psi(\mathbf{r})d\mathbf{r}$ is the probability of finding the particle in the volume element $d\mathbf{r}$ at coordinate $\mathbf{r}$, the expectation value of the position operator $\mathbf{r}$ of the particle, which we write as $\langle \mathbf{r} \rangle$, is given by

$$\langle \mathbf{r} \rangle = \int \mathbf{r}\rho(\mathbf{r})d\mathbf{r} = \int \Psi^*(\mathbf{r})\mathbf{r}\Psi(\mathbf{r})d\mathbf{r}. \tag{1.59}$$

$\langle \mathbf{r} \rangle$ is interpreted as the average of the observable $\mathbf{r}$ measured on a larger number of identically and independently prepared systems represented by the wave function $\Psi(\mathbf{r})$.

Similarly, the expectation value of the momentum operator is

$$\langle \mathbf{p} \rangle = \int d\mathbf{r}\Psi^*(\mathbf{r})\left(-i\hbar\nabla\right)\Psi(\mathbf{r}) = -i\hbar \int d\mathbf{r}\Psi^*(\mathbf{r})\nabla\Psi(\mathbf{r}). \tag{1.60}$$

More generally, for an arbitrary operator, $O(\mathbf{r})$, its expectation value is defined by

$$\langle O \rangle = \int \Psi^*(\mathbf{r},t)O(\mathbf{r})\Psi(\mathbf{r},t)d\mathbf{r}. \tag{1.61}$$

If $\Psi(\mathbf{r})$ is not normalized, then

$$\boxed{\langle O \rangle = \frac{\int \Psi^*(\mathbf{r},t)O(\mathbf{r})\Psi(\mathbf{r},t)d\mathbf{r}}{\int \Psi^*(\mathbf{r},t)\Psi(\mathbf{r},t)d\mathbf{r}}.} \tag{1.62}$$

It should be noted that not all operators yield a measurable expectation value. An operator is an observable if it has a pure real expectation value in all allowed quantum states. In other words, an observable is measurable.

1.10 PROBLEMS

1. Estimate the de Broglie wave length of the following object:

   a. A hydrogen molecule at room temperature

   b. A ^{87}Rb atom at 170 nK (In 1995, Eric Cornell, Carl Wieman, and co-workers at JILA created the Bose-Einstein condensation by cooling a dilute vapor of approximately two thousand rubidium-87 atoms below this temperature)

   c. An electron in a hydrogen atom

   For which one a quantum mechanical treatment is necessary? Why?

2. In 1897, Joseph John Thomson discovered electrons as particles with a definite value of e/μ. In 1927, his son, George Paget Thomson found the wave nature of electron by electron diffraction. In J.J. Thomson's experiment, electrons with kinetic energies of 200 eV passed through a pair of plates with 2 cm separation. Explain why he saw no evidence for wave-like behaviour of electrons.

3. The photoelectric effect, by which a metal surface emits electrons when illuminated by visible and ultra-violet light, was discovered by Heinrich Hertz in 1887. But a given surface only emits electrons when the frequency of the illuminated light exceeds a certain threshold. This threshold corresponds to the minimum energy needed to eject an electron from the surface. This minimum energy is called work function of the metal.

   The work function of magnesium (Mg) is 3.68 eV. To find the minimum frequency of light that is needed in order to produce photoelectrons from magnesium.

4. Consider a one-dimensional wave function $\Psi(x,t) = Ae^{ikx-i\omega t} + Be^{-ikx-i\omega t}$, compute its probability density ρ and probability current density j.

5. Conservation of probability: in a one-dimensional system, let $P_{ab}(t)$ be the probability of finding a particle in the range $a < x < b$ at time t,

   a. show that

   $$\frac{d}{dt}P_{ab} = j(a,t) - j(b,t),$$

   where

   $$j(x,t) = \frac{i\hbar}{2\mu}\left(\Psi\partial_x\Psi^* - \Psi^*\partial_x\Psi\right).$$

   b. Discuss the meaning of this equation.

6. Consider a complex potential, $V(r) = V_0(r) - i\tau$ where $V_0(r)$ and τ are real and $\tau > 0$ is time-independent.

   a. Derive the continuity equation for this system.

   b. Show that the total probability decreases exponentially with time.

7. Suppose $\Psi_1(\mathbf{r},t)$ and $\Psi_2(\mathbf{r},t)$ are two square integrable wave functions that satisfy the same Schrödinger equation. Show that their overlap $\int \Psi_1^*\Psi_2 d\mathbf{r}$ is time independent.

8. $\Psi(\mathbf{r})$ is a nondegenerate eigenstate of the Hamiltonian

$$H = -\frac{\hbar^2}{2\mu}\nabla^2 + V(\mathbf{r}),$$

$V(\mathbf{r})$ is real. Show that the probability current density of Ψ is zero.

9. Find the expectation value of the momentum operator $p_x = -i\hbar\partial/\partial x$ in a one-dimensional bound state whose wave function $\Psi(x)$ is square integrable and

a. $\Psi(x)$ is real.

b. $\Psi(x) = \exp(ikx)\psi(x)$ and $\psi(x)$ is real.

2 One-dimensional Eigen-problem

In this chapter, we solve the eigen-problems for a few simple one-dimensional systems to acquire experience in solving the Schrödinger eigen-equation

$$-\frac{\hbar^2}{2\mu}\frac{d^2}{dx^2}\Psi(x) + V(x)\Psi(x) = E\Psi(x),$$
(2.1)

where $V(x)$ is the potential energy. If two or more independent eigenstates have the same eigenenergy E, we call these energy eigenstates degenerate, otherwise nondegenerate.

One-dimensional problems are physically interesting not only because they are simple, but also because a number of more complicated higher-dimensional problems can be reduced to the solutions of equations in one dimension.

2.1 SYMMETRIC POTENTIAL AND PARITY

Let us first discuss the behavior of the Schrödinger equation in a symmetric potential under the operation of reflection through the origin: $x \rightarrow -x$. This is also called *parity operation*. If the potential is symmetric under the spatial reflection, i.e.

$$V(x) = V(-x),$$
(2.2)

the Hamiltonian

$$H = -\frac{\hbar^2}{2\mu}\frac{d^2}{dx^2} + V(x)$$
(2.3)

is unchanged when x is replaced by $-x$, hence invariant under the parity operation. Thus if we change the sign of x in the Schrödinger equation, we have

$$-\frac{\hbar^2}{2\mu}\frac{d^2}{dx^2}\Psi(-x) + V(x)\Psi(-x) = E\Psi(-x).$$
(2.4)

It implies that both $\Psi(x)$ and $\Psi(-x)$ are solutions of the Schrödinger equation with the same eigenenergy E. Two cases may arise depending on whether the eigenvalue E is degenerate or nondegenerate.

1. The eigenstate is nondegenerate:
   If the eigenvalue E is nondegenerate, $\Psi(x)$ differs from $\Psi(-x)$ only by a multiplicative constant

$$\Psi(x) = a\Psi(-x).$$
(2.5)

DOI: 10.1201/9781003174882-2

Changing the sign of x yields

$$\Psi(-x) = a\Psi(x). \tag{2.6}$$

Thus we have

$$\Psi(x) = a^2\Psi(x), \tag{2.7}$$

hence $a^2 = 1$, so that $a = \pm 1$ and

$$\Psi(x) = \pm\Psi(-x). \tag{2.8}$$

This means that the eigenfunction $\Psi(x)$ has a definite parity, being either even ($a = 1$) or odd ($a = -1$) under the reflection operation.

We can also define a parity operator P, which is to convert $\Psi(x)$ to $\Psi(-x)$

$$P\Psi(x) = \Psi(-x). \tag{2.9}$$

Clearly $P^2 = I$, where I is the identity operator (this identity operator is often replaced by 1). Using Eq. (2.5), we have

$$P\Psi(x) = a\Psi(x). \tag{2.10}$$

Thus a is just the eigenvalue of the parity operator.

2. The eigenstate is degenerate:

In this case, $\Psi(x)$ is not necessary to have a definite parity. Nevertheless, we can always use the linear superposition of $\Psi(x)$ and $\Psi(-x)$ to construct an eigenfunction with definite parity. For example, an eigenstate with an even or odd parity can be represented as

$$\Psi_\pm(x) = (1 \pm P)\Psi(x) = \Psi(x) \pm \Psi(-x).$$

Since $P^2 = 1$, it is simple to show that

$$P\Psi_\pm(x) = P(1 \pm P)\Psi(x) = \pm\Psi_\pm(x).$$

Thus for a symmetric potential, an eigenfunction of the Schrödinger equation can be always chosen to have a definite parity. This property can be used to simplify the calculation of the eigenstate since only the eigenfunction in the positive or negative x-axis needs to be evaluated. The eigenfunction of odd parity vanishes at the origin. The eigenfunction of even parity, on the other hand, must have zero slope at the origin.

2.2   FREE PARTICLE

We first consider a simple example of a particle moving in a constant potential. As the force acting on the particle vanishes, the particle is free. Without loss of generality, we set the potential to zero,

$$V(x) = 0. \tag{2.11}$$

The addition of a constant potential merely shifts the eigenenergy of the Schrödinger equation and leaves the eigenfunction unchanged.

The time-independent Schrödinger equation reads

$$-\frac{\hbar^2}{2\mu}\frac{d^2}{dx^2}\Psi(x) = E\Psi(x). \tag{2.12}$$

This equation has two independent plane-wave solutions at a given energy E,

$$\Psi_k(x) = A\exp(ikx), \tag{2.13}$$

where A is a normalization constant, and

$$k = \pm\frac{\sqrt{2\mu E}}{\hbar}. \tag{2.14}$$

These two-fold degenerate energy states are also the eigenfunctions of the momentum operator p_x

$$p_x\Psi_k(x) = -i\hbar\frac{d}{dx}\Psi_k(x) = \hbar k\Psi_k(x). \tag{2.15}$$

Thus $p = \hbar k$ is the momentum of the eigenfunction $\Psi_k(x)$, and k is the wave number. $\Psi_k(x)$ carries a momentum of $\hbar k$.

If $E < 0$, k becomes purely imaginary. In this case, $\Psi(x)$ would increase infinitely at one of the limits $x \to \infty$ or $x \to -\infty$. Thus the eigenenergy must be nonnegative. As all nonnegative E are allowed, the energy spectrum is continuous.

The plane wave solution has the following properties:

1. The probability density corresponding to $\Psi_k(x)$ is

$$\rho_k(x) = |\Psi_k(x)|^2 = |A|^2, \tag{2.16}$$

   independent on the coordinate x. It indicates that the position of the particle in a momentum eigenstate is completely uncertain. In other words, a particle with well defined momentum, namely with vanish variance in the momentum space $\Delta p = 0$, cannot be localized at all in the coordinate space, namely with infinite variance in real space $\Delta x = \infty$.

2. The probability current density of this plane wave is

$$\begin{aligned} j_k(x) &= \frac{\hbar}{2i\mu}\left[\Psi_k^*(x)\frac{d}{dx}\Psi_k(x) - \Psi_k(x)\frac{d}{dx}\Psi_k^*(x)\right] \\ &= \frac{\hbar}{2i\mu}\left[A^*e^{-ikx}ikAe^{ikx} - Ae^{ikx}(-ik)A^*e^{-ikx}\right] \\ &= \frac{\hbar k}{\mu}|A|^2, \end{aligned} \tag{2.17}$$

   which is also x-independent. This current density can be also expressed as

$$j_k(x) = \frac{\hbar k}{\mu}\rho_k = v_k\rho_k, \tag{2.18}$$

   where $v_k = \hbar k/\mu$ is just the velocity of the particle.

3. The energy eigenstate is generally a linear combination of these two degenerate basis states, and can be expressed as

$$\Psi(x) = Ae^{ikx} + Be^{-ikx}, \tag{2.19}$$

where A and B are arbitrary complex constants. The probability density of this wave function is given by

$$\rho(x) = |\Psi(x)|^2 = |A|^2 + |B|^2 + AB^* e^{2ikx} + A^* B e^{-2ikx}. \tag{2.20}$$

The corresponding probability current density is

$$j(x) = \frac{\hbar k}{\mu}\left(|A|^2 - |B|^2\right), \tag{2.21}$$

which is just the sum of the probability current density of the two travelling plane waves with momenta k and $-k$.

Two particular cases are obtained by setting $A = B$ and $A = -B$.

Case I: $A = B$

The eigenfunction becomes

$$\Psi(x) = C\cos kx, \tag{2.22}$$

where $C = 2A$. This is a standing wave which is symmetric under the operation of spatial reflection and hence has an even parity. The nodes of this wave function, at which $\Psi(x)$ vanishes, are located at

$$x = \frac{\pi}{k}\left(n + \frac{1}{2}\right), \quad n = 0, \pm 1, \pm 2, \cdots \tag{2.23}$$

Its probability density is given by

$$\rho(x) = |\Psi(x)|^2 = |C|^2 \cos^2 kx. \tag{2.24}$$

As $\Psi(x)$ is now a real function, the probability current density is absolutely zero

$$j(x) = 0. \tag{2.25}$$

This is due to the cancellation between the probability flux contributed by the plane wave along the positive k direction and that along the negative k direction.

Case II: $A = -B$

The eigenfunction is again a standing wave

$$\Psi(x) = D\sin kx, \tag{2.26}$$

where $D = 2iA$. This eigenfunction is odd under the spatial reflection and thus has an odd parity. Its physical properties are similar to those of the standing wave previously discussed. The nodes are located at

$$x = \frac{n\pi}{k}, \quad n = 0, \pm 1, \pm 2, \cdots \tag{2.27}$$

Again the probability current density of this standing wave vanishes and the probability density equals

$$\rho(x) = |\Psi(x)|^2 = |D|^2 \sin^2 kx. \tag{2.28}$$

2.3   DELTA-FUNCTION NORMALIZATION

The probability density of the plane wave solution is a constant, independent of the spatial coordinate of the particle. This means that the integral of the probability density in the whole space, from negative infinite to positive infinite in the x-axis, is infinite and the wave function cannot be normalized to 1. Thus this kind of wave function is not square integrable. In this case, we can only speak of relative probabilities. In certain cases, it is sufficient to treat this kind of wave functions without considering the normalization. Generally, it is desirable to treat these wave functions on the same footing as the square integrable ones.

A convenient and commonly used normalization scheme is to normalize the wave function using the Dirac delta function. In this scheme, two plane waves are orthonormalized to a delta function

$$\int_{-\infty}^{\infty} dx \Psi_{k'}^*(x) \Psi_k(x) = \delta(k - k'), \tag{2.29}$$

where $\delta(k)$ is the Dirac delta function defined by the formula

$$\boxed{f(x) = \int_{-\infty}^{\infty} dx' f(x') \delta(x - x'),} \tag{2.30}$$

with $f(x)$ an arbitrary function of x.

Since $f(x)$ is an arbitrary function of x, $\delta(x)$ must be zero if $x \neq 0$ and its integral over any interval covering the point $x = 0$ is 1

$$\int_a^b dx \delta(x) = 1, \tag{2.31}$$

where $a < 0$ and $b > 0$. A special case is $a = 0^-$ and $b = 0^+$.

To explicitly determine a delta function, let us consider the Fourier transformation of $f(x)$

$$F(k) = \frac{1}{\sqrt{2\pi}} \int_{-\infty}^{\infty} dx f(x) e^{-ikx}, \tag{2.32}$$

and its inverse transformation

$$f(x) = \frac{1}{\sqrt{2\pi}} \int_{-\infty}^{\infty} dk F(k) e^{ikx}. \tag{2.33}$$

$F(k)$ is the Fourier transform of $f(x)$ and, similarly, $f(x)$ is the inverse Fourier transform of $F(k)$. A function $f(x)$ or $F(k)$ can be expressed as a Fourier transform if and only if above Fourier integrals converge, when both $f(x)$ and $F(k)$ are square integrable,

$$\int_{-\infty}^{\infty} dx |f(x)|^2 \ < \ \infty, \tag{2.34}$$

$$\int_{-\infty}^{\infty} dk |F(k)|^2 \ < \ \infty, \tag{2.35}$$

Substituting Eq. (2.33) into Eq. (2.32), we obtain

$$f(x) = \frac{1}{2\pi} \int_{-\infty}^{\infty} dx' f(x') \int_{-\infty}^{\infty} dk e^{ik(x-x')}. \tag{2.36}$$

From the definition of the delta function, we have

$$\boxed{\delta(x-x') = \frac{1}{2\pi} \int_{-\infty}^{\infty} dk e^{ik(x-x')}.} \tag{2.37}$$

For the plane wave solution of the free particle, if it is δ-function normalized, then the normalization constant should be $A = 1/\sqrt{2\pi}$. Hence

$$\Psi_k(x) = \frac{1}{\sqrt{2\pi}} e^{ikx} \tag{2.38}$$

is delta-function normalized.

2.4 INFINITE SQUARE WELL POTENTIAL

Now we consider the motion of a particle in an infinite deep well of width L in one dimension. The potential $V(x)$ is defined by

$$V(x) = \begin{cases} 0, & 0 < x < L, \\ \infty, & x < 0 \text{ or } x > L. \end{cases} \tag{2.39}$$

As the potential energy is infinite at $x < 0$ or $x > L$, the probability of finding the particle outside the well is zero. Thus we only need to solve the equation inside the well.

A basic assumption of quantum mechanics is that the wave function is continuous in space. Therefore, $\Psi(x)$ must vanish at the surfaces of the walls, namely

$$\Psi(0) = \Psi(L) = 0. \tag{2.40}$$

In many cases, it is also assumed that the derivative of wave function is continuous in a potential smoothly varying in space. However, for an infinite square well, there is an infinite jump in the potential, the derivative of the wave function cannot be continuous (i.e. zero) at the two boundary points. Otherwise the Schrödinger equation just has a trivial solution $\Psi(x) = 0$ even inside the well.

Inside the well, the Schrödinger equation reads

$$-\frac{\hbar^2}{2\mu}\frac{d^2}{dx^2}\Psi(x) = E\Psi(x). \tag{2.41}$$

The solution of this equation satisfying the condition $\Psi(0) = 0$ is just the standing wave

$$\Psi(x) = A\sin kx, \tag{2.42}$$

where A is the normalization constant. The corresponding eigenenergy is

$$E = \frac{\hbar^2 k^2}{2\mu}. \tag{2.43}$$

The value of k is determined by the boundary condition at $x = L$. Since $\Psi(L) = 0$, we find that k is quantized

$$k = \frac{n\pi}{L}, \tag{2.44}$$

where n is a nonzero integer. The two eigenfunctions corresponding to n and $-n$ differ only by a minus sign, thus they are not independent. We only need to keep the solutions with positive n. Thus the eigenenergy is

$$\boxed{E_n = \frac{n^2\pi^2\hbar^2}{2\mu L^2}, \qquad n = 1,2,3,\cdots.} \tag{2.45}$$

Unlike in the classical system, the ground state energy (i.e. the lowest eigenenergy) of the particle is nonzero

$$E_1 = \frac{\pi^2\hbar^2}{2\mu L^2}. \tag{2.46}$$

The existence of this finite zero-point energy is apparently an effect of quantum physics.

From the normalization condition of the wave function

$$\int_0^L dx\,|\Psi(x)|^2 = A^2 \int_0^L dx\sin^2 kx = \frac{LA^2}{2} = 1, \tag{2.47}$$

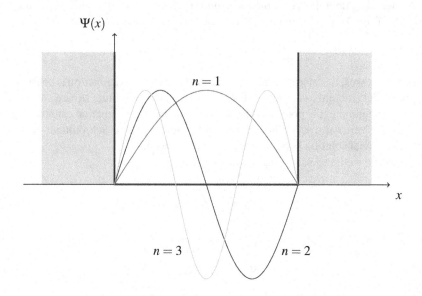

Figure 2.1   The lowest three eigenfunctions of the infinite square well potential.

we find that $A = \sqrt{2/L}$, independent of n. Thus the normalized eigenfunction is

$$\Psi_n(x) = \sqrt{\frac{2}{L}} \sin\frac{n\pi x}{L}, \qquad (0 < x < L). \tag{2.48}$$

The eigensolution of a particle in an infinite square well consists of an infinite number of discrete energy levels. Figure 2.1 shows the lowest three energy eigenfunctions. At each level, there is only one eigen-function. Thus the eigenstate is nondegenerate. The n'th eigenfunction has $(n-1)$ nodes within the well. The two eigenfunctions corresponding to different eigenenergies are orthogonal to each other

$$\int_0^L dx\Psi_n^*(x)\,\Psi_m(x) = \delta_{m,n}. \tag{2.49}$$

2.5 FINITE SQUARE WELL POTENTIAL

We now solve the eigen-equation of a particle in a finite square well potential

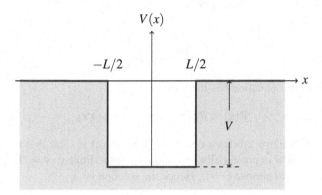

$$V(x) = \begin{cases} -V < 0, & |x| < L/2, \\ 0, & |x| > L/2. \end{cases} \tag{2.50}$$

V is the depth of the potential well and L is its width. This potential is symmetric under the spatial reflection $x \to -x$. In this case, the eigenfunction can be always chosen to have a definite parity. We only need to solve the eigenfunction in the positive x space.

Inside the well, the Schrödinger equation reduces to

$$-\frac{\hbar^2}{2\mu}\frac{d^2}{dx^2}\Psi(x) = (E+V)\Psi(x). \tag{2.51}$$

It can be also expressed as

$$\frac{d^2}{dx^2}\Psi(x) + k^2\Psi(x) = 0, \tag{2.52}$$

where

$$k = \frac{\sqrt{2\mu(E+V)}}{\hbar}. \tag{2.53}$$

To solve the Schrödinger equation including the eigenfunction outside well, two cases must be distinguished, corresponding to the case the energy E is smaller than or larger than 0, respectively. When the energy is less than 0, the particle is confined and hence in a bound state. When the energy is higher than 0, the particle is unconfined and in a scattering state.

2.5.1 BOUND STATES $-V < E \leq 0$

Outside the well, the Schrödinger equation is

$$-\frac{\hbar^2}{2\mu}\frac{d^2}{dx^2}\Psi(x) - E\Psi(x) = 0. \tag{2.54}$$

It can be also expressed as

$$\frac{d^2}{dx^2}\Psi(x) - \kappa^2\Psi(x) = 0,$$

(2.55)

where

$$\kappa = \frac{\sqrt{-2\mu E}}{\hbar} = \sqrt{\frac{2\mu V}{\hbar^2} - k^2}.$$

(2.56)

The solution of this equation has the form

$$\Psi(x) = B\exp(-\kappa x) + C\exp(\kappa x),$$

(2.57)

where B and C are two arbitrary constants. If $C \neq 0$, it is clear that the eigenfunction $\Psi(x)$ will grow exponentially and diverges in the limit $x \to \infty$. To prevent this divergence, we must choose $C = 0$. Hence the solution is

$$\boxed{\Psi(x) = B\exp(-\kappa x), \qquad x > L/2.}$$

(2.58)

$\Psi(x)$ can be taken as a parity eigenstate. Below we consider the even and odd parity solutions separately.

Even parity states

In this case, the solution of Eq. (2.52) is given by

$$\boxed{\Psi(x) = A\cos kx, \qquad x < L/2,}$$

(2.59)

where A is a normalization constant. From the continuity of the wave function and its derivative at the boundary of the well, we obtain the following two equations

$$A\cos\frac{kL}{2} = Be^{-\kappa L/2},$$

(2.60)

$$kA\sin\frac{kL}{2} = \kappa Be^{-\kappa L/2}.$$

(2.61)

Taking the ratio between the above two equations, we further obtain the equation for determining the energy

$$k\tan\frac{kL}{2} = \kappa.$$

(2.62)

Since both k and κ are positive, this requires that

$$n\pi < \frac{Lk}{2} < (n+\frac{1}{2})\pi, \qquad n = 0,1,2...$$

(2.63)

Eliminating κ in Eq. (2.62) using the expression (2.56), we obtain the equation determining k

$$k = \frac{\sqrt{2\mu V}}{\hbar}\left|\cos\frac{kL}{2}\right|.$$

(2.64)

Odd parity states

Now the solution of Eq. (2.52) becomes

$$\boxed{\Psi(x) = A\sin kx, \qquad x < L/2.}$$ (2.65)

By requiring both the wave function and its derivative to be continuous at the boundary of the well, we obtain the following two equations

$$A\sin\frac{kL}{2} = Be^{-\kappa L/2},$$ (2.66)

$$kA\cos\frac{kL}{2} = -\kappa Be^{-\kappa L/2}.$$ (2.67)

Solving these two equations, we find the equation for determining the eigenenergy to be

$$k\cot\frac{kL}{2} = -\kappa.$$ (2.68)

As both k and κ are positive numbers, this equation has solution only when the following condition is satisfied

$$n\pi - \frac{\pi}{2} \le \frac{kL}{2} < n\pi, \qquad (n = 1, 2, \dots).$$ (2.69)

After eliminating κ using Eq. (2.56), Eq. (2.68) becomes

$$k = \frac{\sqrt{2\mu V}}{\hbar}\left|\sin\frac{kL}{2}\right|.$$ (2.70)

Graphical solution

Setting $\zeta = kL/2$ and $V_0 = \sqrt{\mu V L^2/2\hbar^2}$, the equations for determining the eigenenergy can be repressed as

$$
\begin{aligned}
\zeta &= V_0|\cos\zeta|, & n\pi < \zeta < \left(n+\tfrac{1}{2}\right)\pi, & \quad \text{(even parity)}, \\
\zeta &= V_0|\sin\zeta|, & \left(n+\tfrac{1}{2}\right)\pi < \zeta < (n+1)\pi, & \quad \text{(odd parity)}.
\end{aligned}
$$ (2.71)

The solutions of ζ satisfying the above equations are determined by the intersecting points between the curve $V_0|\cos(kL/2)|$ and the straight line ζ for the even parity states, and those between the curve $V_0|\sin(kL/2)|$ and the straight line ζ for the odd parity states, as illustrated in Fig. 2.2. Having obtaining the eigenenergy, the normalization constants, A and B, can be determined from Eq. (2.60) and the normalization condition of the wave function.

From Fig. 2.2, we immediately read out the following information on the eigenstates:

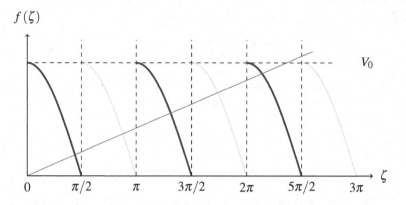

Figure 2.2 Solutions of ζ for the Schrödinger equation with the square potential well are determined by the intersecting points between $f(\zeta) = V_0 |\cos(\zeta)|$ (even parity, thick curve) or $f(\zeta) = V_0 |\sin(\zeta)|$ (odd parity, thin curve) and $f(\zeta) = \zeta$ (solid straight line).

1. All bound states are nondegenerate.
2. There is at least one bound state in the even parity sector no matter how shallow (V) and narrow (L) the potential well is, provided both V and L are finite.
3. A zero-energy bound state, i.e. $E - V = 0^-$, emerges when $\kappa = 0$ or when ζ reaches the value of V_0 at $n\pi/2$ (n positive integer), namely when

$$\zeta = V_0 = \frac{n\pi}{2}, \qquad (2.72)$$

   By further increasing V, the binding energy becomes larger.
4. The total number of bound states increases with the increase of V and equals n if

$$\frac{(n-1)\pi}{2} < V_0 < \frac{n\pi}{2}. \qquad (2.73)$$

   The bound state spectrum consists of alternating even- and odd-parity states. The ground state is always even, the next state is odd, and so on.
5. The n'th eigenfunction has $(n-1)$ nodes at which the wave function becomes zero. The ground state has no nodes. The even parity states have even number of nodes, and the odd parity states have odd number of nodes.
6. The eigenfunction can extend to the classically forbidden region ($|x| > L/2$).

The limit $L \to 0$

Now let us consider a limit in which $L \to 0$ and $V \to \infty$, but the area of the potential well remains a constant, i.e. $VL = \alpha$ (α is a positive constant). In this case, $V_0 =$

$\sqrt{\mu \alpha L / 2\hbar^2} \to 0$, there is only one bound state whose energy determined by the equation

$$\zeta = V_0 \cos \zeta \approx V_0 \left(1 - \frac{1}{2} \zeta^2 \right). \tag{2.74}$$

The solution of ζ is

$$\zeta = \frac{\sqrt{1 + 2V_0^2} - 1}{V_0} \approx V_0 \left(1 - \frac{1}{2} V_0^2 \right), \tag{2.75}$$

and the corresponding eigenenergy is

$$E = \frac{\hbar^2 k^2}{2\mu} - V = \frac{2\hbar^2 \zeta^2}{\mu L^2} - V \approx -\frac{\mu \alpha^2}{2\hbar^2}. \tag{2.76}$$

This energy, as discussed in 2.7, is precisely the solution of the bound state in an attractive delta-function potential.

2.5.2 SCATTERING STATES $E > 0$

When $E > 0$, the particle is no longer bounded and the spectrum becomes continuous. Inside the well, the eigen-equation is unchanged. But the eigen-equation outside the well now becomes

$$\frac{d^2}{dx^2} \Psi(x) + \kappa^2 \Psi(x) = 0, \qquad \kappa = \frac{\sqrt{2\mu E}}{\hbar}. \tag{2.77}$$

To obtain the scattering solution, we assume that the particle is incident upon the well from the left so that the wave function outside the well is given by

$$\Psi(x) = \begin{cases} e^{i\kappa x} + r e^{-i\kappa x}, & x < -L/2, \\ t e^{i\kappa x}, & x > L/2. \end{cases} \tag{2.78}$$

Here the coefficient of the incident wave is set to 1. In this convention, r and t represent the reflection and transmission amplitudes, respectively. The $\exp(-i\kappa x)$ term is absent in the right hand side of the well because there is no reflection wave in the limit $x \to \infty$. It should be emphasized that these scattering wave functions are not normalizable, and do not correspond to any particle states.

From Eq. (1.38), we find the probability current densities of the left and right scattering waves to be

$$j(x) = \begin{cases} \dfrac{\hbar \kappa}{\mu} \left(1 - |r|^2 \right), & x < -L/2, \\[2mm] \dfrac{\hbar \kappa}{\mu} |t|^2, & x > L/2. \end{cases} \tag{2.79}$$

The conservation of probability current density implies that

$$|t|^2 = 1 - |r|^2.$$

(2.80)

This equation holds independent on the detailed shape of the scattering potential.

Inside the well, the wave function can be generally expressed as

$$\Psi(x) = Ae^{ikx} + Be^{-ikx}, \qquad |x| < L/2.$$

(2.81)

The coefficients A and B are related to the reflection and transmission amplitudes, r and t, by the continuity of the wave function and its derivative at the two boundaries. This leads to the equations

$$\begin{cases} e^{-i\kappa L/2} + re^{i\kappa L/2} &=& Ae^{-ikL/2} + Be^{ikL/2}, \\ \kappa e^{-i\kappa L/2} - \kappa re^{i\kappa L/2} &=& kAe^{-ikL/2} - kBe^{ikL/2}, \\ te^{i\kappa L/2} &=& Ae^{ikL/2} + Be^{-ikL/2}, \\ \kappa te^{i\kappa L/2} &=& kAe^{ikL/2} - kBe^{-ikL/2}. \end{cases}$$

(2.82)

These equations can be simplified as

$$\begin{cases} (k+\kappa)e^{-i\kappa L/2} + (k-\kappa)re^{i\kappa L/2} &=& 2kAe^{-ikL/2}, \\ (k-\kappa)e^{-i\kappa L/2} + (k+\kappa)re^{i\kappa L/2} &=& 2kBe^{ikL/2}, \\ (k+\kappa)te^{i\kappa L/2} &=& 2kAe^{ikL/2}, \\ (k-\kappa)te^{i\kappa L/2} &=& 2kBe^{-ikL/2}. \end{cases}$$

(2.83)

Eliminating A and B, we obtain the equations for determining the reflection and transmission coefficients

$$e^{-i\kappa L} + \frac{k-\kappa}{k+\kappa}r = te^{-ikL},$$

(2.84)

$$e^{-i\kappa L} + \frac{k+\kappa}{k-\kappa}r = te^{ikL}.$$

(2.85)

Solving these equations, we obtain the expression of the reflection amplitude

$$r = \frac{e^{-i\kappa L}\left(1 - e^{i2kL}\right)\left(k^2 - \kappa^2\right)}{(k-\kappa)^2 e^{2ikL} - (k+\kappa)^2}.$$

(2.86)

The reflection coefficient is therefore given by

$$|r|^2 = \frac{\left(k^2 - \kappa^2\right)^2 \sin^2 kL}{\left(k^2 - \kappa^2\right)^2 \sin^2 kL + 4k^2\kappa^2} = \frac{V^2 \sin^2 kL}{V^2 \sin^2 kL + 4E(E+V)}.$$

(2.87)

In obtaining the above expression, the following equations are used

$$k^2 - \kappa^2 = \frac{2\mu V}{\hbar^2},$$

(2.88)

$$k\kappa = \frac{2\mu}{\hbar^2}\sqrt{E(E+V)}.$$

(2.89)

The corresponding transmission coefficient is

$$|t|^2 = 1 - |r|^2 = \frac{4E(E+V)}{V^2 \sin^2 kL + 4E(E+V)}. \tag{2.90}$$

It has the following properties

1. The transmission coefficient equals 1 when $\sin kL = 0$, i.e. $k = n\pi/L$. Hence the square well potential is transparent to the incident wave when $k = n\pi/L$. At other k (or energy) points, $|t|^2 < 1$. This is different than in the classical case where the wave is always transmitted.
2. $|t|^2$ reaches a local minimum at $k = (n+1/2)\pi/L$ where $\sin kL = \pm 1$.
3. In the limit $E \ll V$, $|t|^2$ varies linearly with E and vanishes when $E \to 0$:

$$|t|^2 = \frac{4u(1+u)}{\sin^2\left(2V_0\sqrt{1+u}\right) + 4u(1+u)} \longrightarrow \frac{4u}{\sin^2(2V_0)}, \tag{2.91}$$

where $u = E/V$.

2.6 QUANTUM TUNNELING

In classical mechanics, a particle cannot penetrate a potential barrier whose height is higher than the energy of the particle. However, in quantum mechanics, there is a finite probability for a particle to penetrate a barrier even if its potential is higher than the energy of the particle. To understand this quantum tunneling effect, let us study the scattering problem of a particle in a rectangular barrier of height V and width L:

$$V(x) = \begin{cases} V > 0, & 0 < x < L \\ 0 & x < 0, \text{ or } x > L \end{cases}. \tag{2.92}$$

We assume that the particle is incident upon the barrier from the left, and there is no left-going wave on the right hand side of the barrier. In this case, the general solution of the Schrödinger equation outside the barrier is

$$\Psi(x) = \begin{cases} e^{ikx} + re^{-ikx} & x < 0 \\ te^{ikx} & x > L \end{cases}, \tag{2.93}$$

where

$$k = \frac{\sqrt{2\mu E}}{\hbar}. \tag{2.94}$$

r and t are the reflection and transmission amplitudes of the wave, respectively. The corresponding probability current density is

$$j(x) = \begin{cases} v(1 - |r|^2) & x < 0 \\ v|t|^2 & x > L \end{cases}, \tag{2.95}$$

where $v = \hbar k/\mu$ is the velocity. Again, from the conservation of probability current density, we have

$$1 - |r|^2 = |t|^2. \tag{2.96}$$

Inside the barrier, the solution depends on whether the energy $E < V$ or $E > V$. Below we only consider the case $E < V$. The solution is given by

$$\Psi(x) = Ae^{\kappa x} + Be^{-\kappa x}, \qquad 0 < x < L, \tag{2.97}$$

where

$$\kappa = \frac{\sqrt{2\mu (V - E)}}{\hbar}. \tag{2.98}$$

The coefficients A and B, together with r and t, are determined by the continuous conditions of the wave function and its derivative at $x = 0$ and L

$$1 + r = A + B, \tag{2.99}$$
$$ik(1 - r) = \kappa(A - B), \tag{2.100}$$
$$te^{ikL} = Ae^{\kappa L} + Be^{-\kappa L}, \tag{2.101}$$
$$ikte^{ikL} = \kappa \left(Ae^{\kappa L} - Be^{-\kappa L} \right). \tag{2.102}$$

Eliminating A and B, we find that

$$\boxed{\begin{aligned} r &= \frac{\left(k^2 + \kappa^2\right)\left(e^{2\kappa L} - 1\right)}{e^{2\kappa L}\left(k + i\kappa\right)^2 - \left(k - i\kappa\right)^2}, \\[2mm] t &= \frac{4ik\kappa e^{-ikL}e^{\kappa L}}{e^{2\kappa L}\left(k + i\kappa\right)^2 - \left(k - i\kappa\right)^2}. \end{aligned}} \tag{2.103}$$

The reflection and transmission coefficients are given by

$$|r|^2 = \frac{V^2 \sinh^2(\kappa L)}{4E(V - E) + V^2 \sinh^2(\kappa L)}, \tag{2.104}$$

$$|t|^2 = \frac{4E(V - E)}{4E(V - E) + V^2 \sinh^2(\kappa L)}. \tag{2.105}$$

Both the reflection and transmission coefficients are finite. Thus the particle has a finite probability of penetrating through the barrier, which is completely forbidden in classical mechanics. This is known as barrier penetration or tunneling effect.

The transmission coefficient becomes zero when $E \rightarrow 0$. In the limit $E \rightarrow V$

$$|t|^2 \approx \left[1 + \frac{V^2 (\kappa L)^2}{4E (V - E)} \right]^{-1} = \left(1 + \frac{\mu L^2 V}{2\hbar^2} \right)^{-1}. \tag{2.106}$$

It depends on both the width and the height of the barrier, and vanishes in the limit either the barrier width L or the barrier height V going to infinite.

2.7 DELTA-FUNCTION POTENTIAL

The delta-function potential well (or barrier) is a limiting case of the finite potential well (or barrier), which is obtained if the height of the potential becomes infinitely high and the width becomes infinitesimally narrow, while maintaining their product a constant. This is an artificial potential, but it is sufficiently simple to be analytically solved. This potential can be used to simulate a particle which is free to move in two regions separated by a sharp barrier of potential. For example, an electron can move almost freely in a metal, but if two different kinds of metals are put close together, the interface between them acts as a barrier that can be modelled by a delta potential.

The Schrödinger equation for the delta-function potential reads

$$-\frac{\hbar^2}{2\mu} \frac{d^2}{dx^2} \Psi + \alpha \delta(x) \Psi = E \Psi. \tag{2.107}$$

where α is a constant. $\alpha < 0$ corresponds to a potential well and $\alpha > 0$ a potential barrier.

Again we should first solve the equation in the two regions separated by the delta potential independently, and stitch the wave functions in these two regions by imposing the appropriate boundary conditions. If the potential is finite, both the wave function and its derivative should vary smoothly at the boundary points. On the other hand, if the potential becomes infinite, such as in the infinite potential well discussed in Section 2.4, we can still impose the continuous condition of the wave function, but the first derivative of the wave function becomes discontinuous on the boundary. To determine the boundary condition for the derivative of the wave function, let us integrate the Schrödinger equation (2.107) from $-\varepsilon$ to ε in the limit $\varepsilon \rightarrow 0$

$$-\frac{\hbar^2}{2\mu} \int_{-\varepsilon}^{\varepsilon} dx \frac{d^2 \Psi}{dx^2} + \alpha \int_{-\varepsilon}^{\varepsilon} dx \delta(x) \Psi = E \int_{-\varepsilon}^{\varepsilon} dx \Psi. \tag{2.108}$$

On the left hand side of the equation, the first integral is nothing but the first derivative of the wave function, evaluated at the end points, the second integral is just the wave function at the origin. The integral on the right hand side is zero in the limit $\varepsilon \rightarrow 0$. Thus we have

$$\boxed{\frac{\hbar^2}{2\mu} \left(\left. \frac{d\Psi}{dx} \right|_{0-} - \left. \frac{d\Psi}{dx} \right|_{0+} \right) + \alpha \Psi(0) = 0.} \tag{2.109}$$

In a potential barrier, $\alpha > 0$, there is only scattering solution. However, in a potential well, $\alpha < 0$, there may have both bound state ($E < 0$) and scattering state ($E > 0$) solutions. We consider these two kinds of solutions separately.

2.7.1 BOUND STATE ($\alpha < 0$ AND $E < 0$)

When $E < 0$, the general solution to Eq. (2.107) is

$$\Psi(x) = Ae^{\kappa x} + Be^{-\kappa x}, \tag{2.110}$$

where

$$\kappa = \frac{\sqrt{-2\mu E}}{\hbar}. \tag{2.111}$$

In the region $x < 0$, the second term tends to infinite as $x \to -\infty$, so we must choose $B = 0$ and

$$\Psi(x) = Ae^{\kappa x}, \qquad (x < 0). \tag{2.112}$$

In the region $x > 0$, the first term blows up in the limit $x \to \infty$, so we must take $A = 0$ and

$$\Psi(x) = Be^{-\kappa x}, \qquad (x > 0). \tag{2.113}$$

The continuity of the wave function at the origin tells us that $B = A$. By further substituting the above wave functions into Eq. (2.109), we find the equation to determine κ

$$\frac{\hbar^2 \kappa}{\mu} A = -\alpha A. \tag{2.114}$$

Eliminating A from both sides, κ is found to be

$$\kappa = -\frac{\alpha \mu}{\hbar^2}. \tag{2.115}$$

The corresponding energy is

$$E = -\frac{\hbar^2 \kappa^2}{2\mu} = -\frac{\mu \alpha^2}{2\hbar^2}. \tag{2.116}$$

From the normalization of the wave function, we can fix the coefficient A

$$\int_{-\infty}^{\infty} dx |\Psi(x)|^2 = \frac{A^2}{\kappa} = 1, \tag{2.117}$$

so from the normalization of the wave function, we can fix the coefficient A

$$A = \sqrt{\kappa}. \tag{2.118}$$

The above discussion indicates that the delta-function potential well, regardless its strength $|\alpha|$, has one and only one bound state. The probability of the particle at the origin is κ. A schematic plot of the ground state wave function is depicted in Fig. 2.3.

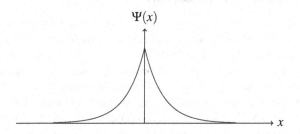

Figure 2.3 Ground state wave function in an attractive δ-function potential.

2.7.2 SCATTERING STATE ($E > 0$)

We now consider the scattering state. Again we assume that the incident wave is from the left hand side of the potential, so that the solution is

$$\Psi(x) = \begin{cases} e^{ikx} + re^{-ikx}, & x < 0, \\ te^{ikx}, & x > 0, \end{cases} \tag{2.119}$$

where

$$k = \frac{\sqrt{2\mu E}}{\hbar}. \tag{2.120}$$

From the continuous condition of the wave function at the origin and Eq. (2.109), we obtain the equations for determining r and t

$$1 + r = t, \tag{2.121}$$

$$\frac{ik\hbar^2}{2\mu}(1 - r - t) + \alpha t = 0. \tag{2.122}$$

The solution is

$$r = -\frac{ib}{1 + ib}, \tag{2.123}$$

$$t = \frac{1}{1 + ib}, \tag{2.124}$$

where $b = \alpha\mu/(k\hbar^2)$. Therefore, the reflection and transmission coefficients of the incident wave are

$$|r|^2 = \frac{b^2}{1 + b^2} = \frac{1}{1 + \left(2\hbar^2 E/\mu\alpha^2\right)}, \tag{2.125}$$

$$|t|^2 = \frac{1}{1 + b^2} = \frac{1}{1 + \left(\mu\alpha^2/2\hbar^2 E\right)}. \tag{2.126}$$

The result indicates that the higher the energy, the greater the probability of transmission. Moreover, the transmission coefficient does not depend on the sign of α.

2.8 THE WKB APPROXIMATION

The WKB method is a technique for finding approximate solutions to linear differential equations with spatially varying coefficients, including the one-dimensional time-independent Schrödinger equation. This method is named after physicists G. Wentzel, H. A. Kramers, and L. Brillouin who independently introduced it in quantum mechanics. It is particularly useful in evaluating bound state energies and tunneling probabilities through potential barriers in quantum mechanics.

To understand the WKB method, let us consider the one-dimensional Schrödinger equation (2.1) for a particle moving in a slowly varying potential $V(x)$. If the potential $V(x) = V_0$ does not depend on x, the solution would be written as a linear superposition of the plane waves

$$\Psi(x) = A \exp(\pm i k_0 x), \tag{2.127}$$

where

$$k_0 = \frac{\sqrt{2\mu(E - V_0)}}{\hbar} \tag{2.128}$$

is the characteristic wave vector of the particle. A is a constant.

The assumption of slowly varying potential means that $V(x)$ only changes slightly over the de Broglie wavelength

$$\lambda(x) = \frac{2\pi}{k(x)} \tag{2.129}$$

for a particle whose energy $E > V(x)$. $k(x)$ is the wave vector at the point x

$$k(x) = \frac{\sqrt{2\mu[E - V(x)]}}{\hbar}. \tag{2.130}$$

In the classical limit, the de Broglie wavelength $\lambda(x)$ tends to zero. Therefore, the slow variance of the potential implies that the WKB approximation is a semiclassical approach. This motivates us to seek the solution of the form

$$\Psi(x) = A \exp[iK(x)], \tag{2.131}$$

where $K(x)$ plays the role of dimensionless classical action.

Substituting (2.131) into (2.1), we obtain the equation for $K(x)$

$$i\frac{d^2 K(x)}{dx^2} - \left[\frac{dK(x)}{dx}\right]^2 + \frac{2\mu[E - V(x)]}{\hbar^2} = 0. \tag{2.132}$$

This is a nonlinear equation which is not easier to solve than the original Schrödinger equation. However, the first term in the above equation is proportional to $\hbar$ which vanishes in the classical limit. This suggests that we can treat $\hbar$ as a small parameter and expand $K(x)$ in the power series

$$\boxed{K(x) = \frac{1}{\hbar}K_0(x) + K_1(x) + \hbar K_2(x) + \cdots.} \tag{2.133}$$

Inserting it into (2.132) and equating to zero the coefficient of each power of $\hbar$ separately, we obtain the equations for the first two terms

$$-\left[\frac{dK_0(x)}{dx}\right]^2 + 2\mu\left[E - V(x)\right] = 0, \tag{2.134}$$

$$i\frac{d^2 K_0(x)}{dx^2} - 2\frac{dK_0(x)}{dx}\frac{dK_1(x)}{dx} = 0. \tag{2.135}$$

Depending on the value of the potential with respect to E, i.e. $V(x) - E$, we can divide the coordinates into two types of regions. One is the classically allowed region with $E > V(x)$, and the other is the classically forbidden region with $E < V(x)$. A position where the potential energy $V(x)$ equals the total energy E is called a turning point. For example, for the following potential well

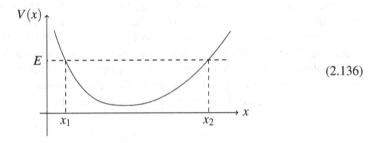

$$\tag{2.136}$$

x_1 and x_2 are the two turning points. The region between these two turning points are classically allowed.

We first discuss the solutions of Eqs. (2.134) and (2.135) in these two types of regions.

1. Region $E > V(x)$:
   The solution for the above equations is given by

   $$K_0(x) = \pm\int^x dx'\sqrt{2\mu\left[E - V(x')\right]}, \tag{2.137}$$

   $$K_1(x) = \frac{i}{2}\ln\sqrt{2\mu\left[E - V(x)\right]} + a, \tag{2.138}$$

   where a is an arbitrary constant. Thus up to the leading two terms, the solution is

   $$\Psi(x) = \frac{1}{\sqrt{k(x)}}\left[Ae^{i\int^x dx'k(x')} + Be^{-i\int^x dx'k(x')}\right], \tag{2.139}$$

   A and B are two arbitrary constants, and $k(x)$ is the wave vector defined by Eq. (2.130).
2. Region $E < V(x)$:

Similarly, we can find the WKB solution in this classically forbidden region

$$\Psi(x) = \frac{1}{\sqrt{\kappa(x)}} \left[Ce^{\int^x dr\kappa(x')} + De^{-\int^x dx'\kappa(x')} \right], \qquad (2.140)$$

where

$$\kappa(x) = \frac{\sqrt{2\mu[V(x) - E]}}{\hbar}. \qquad (2.141)$$

C and D are arbitrary constants.

2.8.1 SOLUTION AROUND A TURNING POINT

Apparently, the WKB approximation is not valid at the turning points where the de Broglie wavelengthes diverge. This suggests that the two WKB wave functions above obtained, (2.139) and (2.140), are not valid in the vicinity of each turning point. Nevertheless, as the wave function is continuous and smooth in the whole space, we should be able to find a patching wave function to join the two types of WKB solutions across a turning point. As the patching wave function is needed just in a small neighborhood of a turning point, say the right turning point x_2 in (2.136), we are allowed to approximate the potential by a straight line around this point

$$V(x) \approx E + V'(x_2)(x - x_2). \qquad (2.142)$$

Thus around x_2, the Schrödinger equation becomes

$$\frac{d^2\Psi(x)}{dx^2} = \alpha^3(x - x_2)\Psi(x), \qquad (2.143)$$

where

$$\alpha = \left[\frac{2\mu V'(x_2)}{\hbar^2} \right]^{1/3}. \qquad (2.144)$$

By absorbing α into the coordinate variable

$$y = \alpha(x - x_2), \qquad (2.145)$$

Eq. (2.143) becomes the standard Airy equation (or the Stokes equation)

$$\frac{d^2\Psi(y)}{dy^2} = y\Psi(y). \qquad (2.146)$$

This is the simplest second-order linear differential equation with a turning point where the character of the solutions changes from oscillatory to exponential. Two linearly independent solutions to this differential equation are the Airy functions $Ai(y)$ and $Bi(y)$, named after the British astronomer G. B. Airy.

The integral representation of the Airy functions is

$$Ai(y) = \frac{1}{\pi} \int_0^\infty dt \cos\left(\frac{t^3}{3} + yt\right), \tag{2.147}$$

$$Bi(y) = \frac{1}{\pi} \int_0^\infty dt \left[\exp\left(-\frac{t^3}{3} + yt\right) + \sin\left(\frac{t^3}{3} + yt\right)\right]. \tag{2.148}$$

In the limit $y \gg 0$, they have the following asymptotic forms

$$Ai(y) \approx \frac{1}{2\sqrt{\pi}y^{1/4}} \exp\left(-\frac{2}{3}y^{3/2}\right), \tag{2.149}$$

$$Bi(y) \approx \frac{1}{\sqrt{\pi}y^{1/4}} \exp\left(\frac{2}{3}y^{3/2}\right). \tag{2.150}$$

In the limit $y \ll 0$, on the other hand, the asymptotic forms become

$$Ai(y) \approx \frac{1}{\sqrt{\pi}(-y)^{1/4}} \sin\left[\frac{2}{3}(-y)^{3/2} + \frac{\pi}{4}\right], \tag{2.151}$$

$$Bi(y) \approx \frac{1}{\sqrt{\pi}(-y)^{1/4}} \cos\left[\frac{2}{3}(-y)^{3/2} + \frac{\pi}{4}\right]. \tag{2.152}$$

In the patching region, the wave function is a linear superposition of $Ai(y)$ and $Bi(y)$

$$\Psi(y) = a\,Ai(y) + b\,Bi(y), \tag{2.153}$$

a and b are constants.

2.8.2 THE CONNECTION FORMULAE

Now let us consider how to splice the WKB wave functions using the patching wave function at a turning point. We first take the right turning point in (2.136) as an example to elaborate the method.

1. The classically forbidden region $x > x_2$:
   In this region, we need only keep the solution that decays exponentially at $x \to \infty$. In this case, the WKB solution is

$$\Psi(x) = \frac{D}{\sqrt{\kappa(x)}} e^{-\int_{x_2}^x dx' \kappa(x')}, \qquad (x > x_2). \tag{2.154}$$

In the vicinity of x_2, Eq. (2.142) holds,

$$\kappa(x) \approx \frac{\sqrt{2\mu V'(x_2)(x - x_2)}}{\hbar} = \alpha^{3/2}\sqrt{x - x_2}, \tag{2.155}$$

and

$$\int_{x_2}^x dx' \kappa(x') \approx \frac{2}{3}\alpha^{3/2}(x - x_2)^{3/2}, \tag{2.156}$$

so that

$$\Psi(x) \approx \frac{D}{\alpha^{3/4}(x - x_2)^{1/4}} e^{-\frac{2}{3}\alpha^{3/2}(x - x_2)^{3/2}}. \qquad (2.157)$$

This expression holds when the relative derivation of the potential at x with respect to the linear potential is a small number, i.e.

$$\varepsilon = \frac{V''(x_2)(x - x_2)^2}{2V'(x_2)(x - x_2)} = \frac{V''(x_2)(x - x_2)}{2V'(x_2)} \ll 1. \qquad (2.158)$$

This suggests that we can choose a small region around a point x which is sufficiently close to the turning point x_2 such that the above inequality holds and yet far enough away from the turning point that the WKB approximation is reliable and

$$y = \alpha(x - x_2) = \frac{2\alpha\varepsilon V'(x_2)}{V''(x_2)} \gg 1. \qquad (2.159)$$

In this case, the patching wave function is asymptotically given by

$$\Psi(y) = \frac{a}{2\sqrt{\pi}z^{1/4}} \exp\left(-\frac{2}{3}z^{3/2}\right) + \frac{b}{\sqrt{\pi}z^{1/4}} \exp\left(\frac{2}{3}z^{3/2}\right). \qquad (2.160)$$

Comparing with Eq. (2.165), we find that

$$a = \sqrt{\frac{4\pi}{\alpha}}D, \qquad b = 0. \qquad (2.161)$$

2. The classically allowed region $x < x_2$:
   The WKB wave function is

$$\Psi(x) = \frac{1}{\sqrt{k(x)}}\left[Ae^{i\int_{x_2}^{x} dx' k(x')} + Be^{-i\int_{x_2}^{x} dx' k(x')}\right], \qquad (2.162)$$

In the vicinity of x_2, as

$$k(x) \approx \frac{\sqrt{2\mu V'(x_2)(x_2 - x)}}{\hbar} = \alpha^{3/2}\sqrt{x_2 - x}, \qquad (2.163)$$

and

$$\int_{x_2}^{x} dx' k(x') \approx -\frac{2}{3}\alpha^{3/2}(x_2 - x)^{3/2}, \qquad (2.164)$$

so we have

$$\Psi(x) = \frac{1}{\alpha^{3/4}(x_2 - x)^{1/4}}\left[Ae^{-i\frac{2}{3}\alpha^{3/2}(x_2 - x)^{3/2}} + Be^{i\frac{2}{3}\alpha^{3/2}(x_2 - x)^{3/2}}\right]. \qquad (2.165)$$

Meanwhile, the patching wave function in the large negative y limit is asymptotically given by

$$\Psi(x) = \frac{a}{\sqrt{\pi}\alpha^{1/4}(x_2 - x)^{1/4}} \sin\left[\frac{2}{3}\alpha^{3/2}(x_2 - x)^{3/2} + \frac{\pi}{4}\right]. \qquad (2.166)$$

Comparing it with the WKB wave function, we find

$$B = -i\frac{\sqrt{\alpha}}{\sqrt{4\pi}}ae^{i\pi/4} = -iDe^{i\pi/4}, \qquad A = iDe^{-i\pi/4}. \qquad (2.167)$$

These are the connection formulas, bridging the WKB solutions on the two sides of the turning point.

The above derivation allows us to express the WKB wave functions on the two sides of x_2 in terms of just one normalization constant D

$$\Psi(x) = \begin{cases} \dfrac{2D}{\sqrt{k(x)}} \sin\left[\dfrac{\pi}{4} + \displaystyle\int_x^{x_2} dx k(x)\right], & (x < x_2), \\[4mm] \dfrac{D}{\sqrt{\kappa(x)}} \exp\left[-\displaystyle\int_{x_2}^x dx \kappa(x)\right], & (x > x_2). \end{cases} \qquad (2.168)$$

Repeating the above procedure for the downward-sloping turning point x_1 in (2.130), we find that the WKB wave function is also parameterized by one normalization constant D'

$$\Psi(x) = \begin{cases} \dfrac{2D'}{\sqrt{k(x)}} \sin\left[\dfrac{\pi}{4} + \displaystyle\int_{x_1}^x dx k(x)\right], & (x > x_1), \\[4mm] \dfrac{D'}{\sqrt{\kappa(x)}} \exp\left[-\displaystyle\int_x^{x_1} dx \kappa(x)\right], & (x < x_1). \end{cases} \qquad (2.169)$$

2.8.3 QUANTIZATION OF ENERGY LEVELS

For the potential illustrated in (2.136) with just two turning points at x_1 and x_2, the WKB wave functions inside the well obtained from Eqs. (2.168) and (2.169) should be equal to each other

$$\frac{2D}{\sqrt{k(x)}} \sin\left[\frac{\pi}{4} + \int_x^{x_2} dx k(x)\right] = -\frac{2D'}{\sqrt{k(x)}} \sin\left[-\frac{\pi}{4} + \int_x^{x_1} dx k(x)\right] \qquad (2.170)$$

The arguments of the sine functions must be equal with modulo π at all the points x. This yields the energy quantization condition

$$\frac{\pi}{4} + \int_x^{x_2} dx k(x) = -\frac{\pi}{4} + \int_x^{x_1} dx k(x) + n\pi, \qquad (2.171)$$

or

$$\int_{x_1}^{x_2} dx k(x) = \left(n - \frac{1}{2}\right)\pi, \qquad (2.172)$$

n is an integer.

This quantization condition determines the energy levels allowed for a potential well with two turning points. The application of this condition requires n to be large and the separation between the two turning points is much larger than the de Broglie wavelength so that the WKB approximation holds. In many cases, the WKB method also works quite well even for small n.

2.9 PROBLEMS

1. Solve the eigenvalue equation

$$-i\hbar \frac{\partial}{\partial \phi} \Psi(\phi) = \lambda \Psi(\phi)$$

subject to the boundary condition $\Psi(\phi + 2\pi) = \Psi(\phi)$. Find all possible eigenvalues λ and the corresponding eigenfunctions.

2. A particle of mass μ moving in one dimension is confined in an infinite square well of width L. In addition, the particle experiences a delta-function potential of strength α located at the center of the well. The potential is

$$V(x) = \begin{cases} \alpha \delta(x), & |x| < \dfrac{L}{2}, \\\\ +\infty, & |x| > \dfrac{L}{2}. \end{cases}$$

Find the equation for the energy eigenvalues in terms of the mass, the potential strength α, and the size L of the system.

3. Solve the Schrödinger equation for a particle moving in a half-plane potential

$$V(x) = \begin{cases} +\infty, & x < 0, \\\\ -V < 0, & 0 < x < L, \\\\ 0, & x > L, \end{cases}$$

a. Find the eigen-spectrum of the bound states.

b. Determine the condition that there is only one bound state.

c. For the scattering state, calculate the reflection coefficient.

4. A particle approaches a potential step from the left with an energy $E > V > 0$. The Hamiltonian is defined by

$$H = \begin{cases} -\dfrac{\hbar^2}{2\mu_1} \dfrac{\partial^2}{\partial x^2}, & x < 0, \\\\ -\dfrac{\hbar^2}{2\mu_2} \dfrac{\partial^2}{\partial x^2} + V, & x > 0. \end{cases}$$

a. Find the amplitudes of the transmitted and reflected waves.

b. Find the spatial dependence of the probability current density, and discuss why the probability current densities on the two sides of $x = 0$ are not conserved.

5. Derive the equations that determine the bound state eigenvalues of a particle in a one-dimensional attractive double δ-function potential

$$V(x) = -b [\delta(x-a) + \delta(x+a)].$$

6. $\Psi(x)$ is a real eigenfunction of bound state, show that the expectation value of the momentum operator $p_x = -i\hbar \partial / \partial x$ vanishes in this state.

7. For the one-dimensional harmonic oscillator potential

$$V(x) = \frac{1}{2}\mu\omega^2 x^2,$$

find the energy levels using the WKB method.

8. Assuming x_1 is a downward-sloping turning point, derive the connection formula at that point and verify Eq. (2.169).

9. *For an arbitrary but smoothly varying potential barrier,

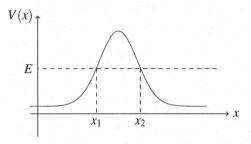

find the transmission probability using the WKB approximation.

3 Representation theory of quantum states

3.1 REPRESENTATION

3.1.1 DIRAC BRACKET NOTATIONS

In the preceding chapters, a quantum state is represented by a wave function in real space. For doing this, we have actually chosen a particular "coordiante" or representation to represent a quantum state. One can in principle to represent the quantum state using a wave function defined in momentum space or any other basis space. Obviously, this is inconvenient to use a wave function defined in a particular basis space for a general discussion.

In his famous monograph "The Principles of Quantum Mechanics", Dirac[1] introduced a convenient notation to represent a state and its expectation values. Using these notations, physical variables and quantum states can be represented independent of the choice of bases (or representation).

In Dirac's notation, an inner product (also called scalar product) of two square integrable functions Ψ_1 and Ψ_2 is denoted by the symbol $\langle \Psi_2 | \Psi_1 \rangle$

$$\langle \Psi_2 | \Psi_1 \rangle = \int \Psi_2^* (\mathbf{r}) \Psi_1 (\mathbf{r}) \, d\mathbf{r}. \tag{3.1}$$

This is a generalization of the inner product of two vectors in linear algebra. The symbol $|\Psi\rangle$ is known as a ket of Ψ while $\langle \Psi |$ is known as a bra of Ψ. The right hand side of this equation is represented as a spatial integration. But it should be emphasized that the expression $\langle \Psi_2 | \Psi_1 \rangle$ can represent an inner product between an initial state $|\Psi_1\rangle$ and a final state $|\Psi_2\rangle$ in an arbitrary representation. Hence it is a more general expression for the inner product.

From the above expression, we find that the complex conjugate of an inner produce is simply to swap the bra with the ket

$$\langle \Psi_2 | \Psi_1 \rangle^* = \langle \Psi_1 | \Psi_2 \rangle. \tag{3.2}$$

The inner product of a wave function with itself

$$\langle \Psi | \Psi \rangle = \int d\mathbf{r} |\Psi (\mathbf{r})|^2 \tag{3.3}$$

---

[1] P. Dirac, The Principles of Quantum Mechanics, Oxford University Press, 1930.

DOI: 10.1201/9781003174882-3

is clearly real and positive if $\Psi(\mathbf{r})$ is not completely zero. Ψ is normalized if its inner product equals 1

$$\langle \Psi | \Psi \rangle = 1. \tag{3.4}$$

Two quantum states, Ψ_1 and Ψ_2, are orthogonal to each other if their inner product equals zero

$$\langle \Psi_2 | \Psi_1 \rangle = \langle \Psi_1 | \Psi_2 \rangle = 0. \tag{3.5}$$

In Dirac's notation, the overlap (or inner product) between $|O\Psi_1\rangle = O|\Psi_1\rangle$ and $|\Psi_2\rangle$ is

$$\langle \Psi_2 | O\Psi_1 \rangle = \langle \Psi_2 | O | \Psi_1 \rangle. \tag{3.6}$$

The expectation value of operator O, defined by Eq. (1.62), is now simply expressed as

$$\boxed{\langle O \rangle = \frac{\langle \Psi | O | \Psi \rangle}{\langle \Psi | \Psi \rangle}.} \tag{3.7}$$

3.1.2 REPRESENTATION OF QUANTUM STATES

A number of quantum states $\{\Psi_n\}$ form an orthonormal and complete basis set if they are normalized and mutually orthogonal,

$$\langle \Psi_m | \Psi_n \rangle = \delta_{m,n}, \tag{3.8}$$

and any quantum state Ψ can be expressed as a linear combination of them, i.e

$$|\Psi\rangle = \sum_n \psi_n |\Psi_n\rangle. \tag{3.9}$$

The coefficient ψ_n is just the wave function represented in the basis space spanned by $\{|\Psi_n\rangle, n = 1, 2, \cdots\}$. Using the orthonormality of the basis states, we find that

$$\boxed{\psi_n = \langle \Psi_n | \Psi \rangle.} \tag{3.10}$$

In Eq. (3.8), $\delta_{m,n}$ is the Kronecker delta symbol defined by

$$\boxed{\delta_{m,n} = \begin{cases} 1, & m = n, \\ \\ 0, & m \neq n. \end{cases}} \tag{3.11}$$

A particular example is the wave function of a particle in the coordinate space, $\Psi(\mathbf{r})$, which is a real-space representation of the state $|\Psi\rangle$, i.e.

$$\Psi(\mathbf{r}) = \langle \mathbf{r} | \Psi \rangle, \tag{3.12}$$

and the quantum state can be expressed as

$$|\Psi\rangle = \int d\mathbf{r} \Psi(\mathbf{r}) |\mathbf{r}\rangle. \tag{3.13}$$

If $|\Psi\rangle = |p\rangle$ is an eigenstate of the momentum operator $p_x = -i\hbar\partial_x$, then

$$|p\rangle = \int_{-\infty}^{\infty} dx \frac{1}{\sqrt{2\pi}} e^{ipx/\hbar} |x\rangle, \qquad (3.14)$$

and

$$
\begin{aligned}
\langle x|p\rangle &= \int_{-\infty}^{\infty} dx' \frac{1}{\sqrt{2\pi}} e^{ipx'/\hbar} \langle x|x'\rangle \\
&= \int_{-\infty}^{\infty} dx' \frac{1}{\sqrt{2\pi}} e^{ipx'/\hbar} \delta(x'-x) \\
&= \frac{1}{\sqrt{2\pi}} e^{ipx/\hbar}
\end{aligned}
\qquad (3.15)
$$

is the wave function of this momentum eigenstate in real space.

If we group all the basis states as a row vector, Eq. (3.9) can then be also expressed as

$$|\Psi\rangle = (|\Psi_1\rangle, |\Psi_2\rangle, \cdots)\,\Psi, \qquad (3.16)$$

where $\Psi = (\psi_1, \psi_2, \cdots)^T$ is a column vector representation of the wave function. Furthermore, if $|\Psi\rangle$ is normalized, then

$$\rho_n = |\langle \Psi_n|\Psi\rangle|^2 = |\psi_n|^2 \qquad (3.17)$$

is just the probability of the basis state $|\Psi_n\rangle$ in $|\Psi\rangle$.

3.1.3 HERMITIAN OPERATORS

Adjoint or Hermitian conjugate of an operator O, written as $O^\dagger$, is the operator generalization of the complex conjugate of a complex number. It is defined by the formula

$$\boxed{\langle \Phi|O^\dagger|\Psi\rangle = \langle O\Phi|\Psi\rangle = \langle \Psi|O|\Phi\rangle^*,} \qquad (3.18)$$

where Ψ and Φ are any pair of square-integrable functions.

If $|\Psi\rangle$ is an eigenstate of A with the corresponding eigenvalue a

$$A|\Psi\rangle = a|\Psi\rangle, \qquad (3.19)$$

then $\langle\Psi|$ is a left eigenstate of $A^\dagger$ with the corresponding eigenvalue a^*

$$\langle\Psi|A^\dagger = a^*\langle\Psi|. \qquad (3.20)$$

The Hermitian conjugate of $|\Psi\rangle$ equals $\langle\Psi|$, i.e.

$$|\Psi\rangle^\dagger = \langle\Psi|. \qquad (3.21)$$

Similarly,

$$(\langle\Psi|)^\dagger = |\Psi\rangle. \qquad (3.22)$$

Operator O is said to be self-adjoint or Hermitian if

$$O^\dagger = O, \tag{3.23}$$

A Hermitian operator is the operator generalization of a real number. All its eigenvalues are real. This is because if $|\Psi\rangle$ is a normalized eigenstate of O

$$O|\Psi\rangle = a|\Psi\rangle \tag{3.24}$$

with a the corresponding eigenvalue, then

$$a = \langle\Psi|O|\Psi\rangle = \langle\Psi|O^\dagger|\Psi\rangle = a^*. \tag{3.25}$$

Any physical observable, such as the momentum, is a self-adjoint or Hermitian operator. To prove the omentum operator $\mathbf{p} = -i\hbar\nabla$ to be a Hermitian operator, let us consider the inner product of the hermitian conjugate of the momentum operator between two arbitrary square integrable wave functions Φ and Ψ. From the defining equation of the Hermitian conjugate, we have

$$
\begin{aligned}
\langle\Phi|\mathbf{p}^\dagger|\Psi\rangle &= \int d\mathbf{r}\,[-i\hbar\nabla\Phi(\mathbf{r})]^*\,\Psi(\mathbf{r}) \\
&= i\hbar\int d\mathbf{r}\,[\nabla\Phi^*(\mathbf{r})]\,\Psi(\mathbf{r}) \\
&= i\hbar\oint_{S_\infty} d\mathbf{r}\Phi^*(\mathbf{r})\Psi(\mathbf{r}) + \int d\mathbf{r}\Phi(\mathbf{r})^*\,[-i\hbar\nabla\Psi(\mathbf{r})] \\
&= \langle\Phi|\mathbf{p}|\Psi\rangle. \tag{3.26}
\end{aligned}
$$

This shows that $\mathbf{p}^\dagger = \mathbf{p}$. The integration of $\Phi^*(\mathbf{r})\Psi(\mathbf{r})$ on the infinity boundary vanishes due to the square integrability of the two wave functions.

For a derivative operator, such as the momentum operator $\mathbf{p} = -i\hbar\nabla$, caution should be taken in performing a hermitian conjugate. For example, $(-i\hbar\nabla\Psi)^\dagger = \Psi^\dagger(-i\hbar\nabla)$ holds as an operator identity. This hermitian conjugate can be also done without changing the order of ∇ and Ψ. In this case, $(\nabla\Psi)^\dagger = \nabla\Psi^\dagger$, and

$$(-i\hbar\nabla\Psi)^\dagger = i\hbar\left(\nabla\Psi^\dagger\right) = \Psi^\dagger(-i\hbar\nabla). \tag{3.27}$$

Similarly, for the time derivative, we have

$$(i\hbar\partial_t\Psi)^\dagger = -i\hbar\left(\partial_t\Psi^\dagger\right) = \Psi^\dagger(i\hbar\partial_t). \tag{3.28}$$

3.1.4 EIGENSTATES OF HERMITIAN OPERATORS

The eigenstates of a Hermitian operator have two fundamental properties:

I. Orthogonality

> Theorem: All the eigenstates, $\{|\Psi_n\rangle, n = 1, 2, \ldots\}$, of a Hermitian operator A
>
> $$A|\Psi_n\rangle = a_n|\Psi_n\rangle, \tag{3.29}$$
>
> can be always orthonormalized, i.e.
>
> $$\langle\Psi_n|\Psi_m\rangle = \delta_{n,m}. \tag{3.30}$$

To prove this theorem, we first consider the case where $|\Psi_n\rangle$ and $|\Psi_m\rangle$ are the eigenstates of A with different eigenvalues, $a_n \neq a_m$. These two states are orthogonal to each other, i.e.

$$\langle\Psi_m|\Psi_n\rangle = 0 \tag{3.31}$$

because

$$
\begin{aligned}
&(a_n - a_m)\langle\Psi_m|\Psi_n\rangle \\
=\ & a_n\langle\Psi_m|\Psi_n\rangle - a_m\langle\Psi_m|\Psi_n\rangle \\
=\ & \langle\Psi_m|A|\Psi_n\rangle - \langle\Psi_m|A|\Psi_n\rangle \\
=\ & 0.
\end{aligned} \tag{3.32}
$$

Here we have used the fact that A is Hermitian. Since $a_n \neq a_m$, we obtain Eq. (3.31).

We then consider the case that an eigenvalue a_n is α-fold degenerate, and $|\Psi_{nr}\rangle$ $(r = 1, \cdots, \alpha)$ are the corresponding eigenstates

$$A|\Psi_{nr}\rangle = a_n|\Psi_{nr}\rangle. \tag{3.33}$$

Since any linear combination of these eigenstates is also eigenstate of A with eigenvale a_n, it is straightforward to orthogonalize them using the Schmidt orthogonalization, so that they are mutually orthogonal and each is normalized to unity

$$\langle\Psi_{nr}|\Psi_{ns}\rangle = \delta_{r,s}. \tag{3.34}$$

It is also simple to normalize a nondegenerate eigenstate.

II. Completeness

For each orthonormalized eigenstate $|\Psi_n\rangle$ of A, $P_n = P_n^\dagger = |\Psi_n\rangle\langle\Psi_n|$ defines a projection operation, which satisfies the property

$$P_n^2 = (|\Psi_n\rangle\langle\Psi_n|)(|\Psi_n\rangle\langle\Psi_n|) = |\Psi_n\rangle(\langle\Psi_n|\Psi_n\rangle)\langle\Psi_n| = |\Psi_n\rangle\langle\Psi_n| = P_n. \tag{3.35}$$

It is to map a quantum state $|\Psi\rangle$ to the state $|\Psi_n\rangle$

$$P_n|\Psi\rangle = |\Psi_n\rangle\langle\Psi_n|\Psi\rangle = \psi_n|\Psi_n\rangle. \tag{3.36}$$

Thus all the eigenstates of A form a complete basis set because any quantum state can be expended using these eigenstates. The completeness of a basis set is defined by summing over all projection operators of A, i.e.

$$\sum_n P_n = \sum_n |\Psi_n\rangle \langle\Psi_n| = I, \tag{3.37}$$

where I is the unit operator that leaves any state unchanged

$$I|\Psi\rangle = |\Psi\rangle. \tag{3.38}$$

Eq. (3.37) holds because

$$\sum_n |(\Psi_n)\langle\Psi_n|)|\Psi\rangle = \sum_n \psi_n|\Psi_n\rangle = |\Psi\rangle \tag{3.39}$$

Hence, the eigenstates of A span a representation of the Hilbert space, from which any states and operators can be expanded.

If the eigen-spectrum of A is continuous, the complete condition becomes

$$\int d\Psi_n |\Psi_n\rangle \langle\Psi_n| = I. \tag{3.40}$$

For example, all the coordinates x in one dimension form a complete basis set

$$\int dx |x\rangle \langle x| = I. \tag{3.41}$$

On the other hand, if A contains both discrete and continuous eigenstates, then the complete condition becomes

$$\sum_n |\Psi_n\rangle \langle\Psi_n| + \int d\Psi_n |\Psi_n\rangle \langle\Psi_n| = I. \tag{3.42}$$

The energy eigenstates of a particle moving in a finite square well potential discussed in Section 2.5, for example, have a spectrum that contains both discretized bound states and continuous scattering states).

In the discussion below, if no particular specification, I will simply use $\sum_n$ to represent the summation for discrete eigenstates as well as the integration for continuous eigenstates, namely

$$\sum_n \longleftrightarrow \sum_n + \int d\Psi_n. \tag{3.43}$$

3.1.5 REPRESENTATION OF OPERATORS

An operator O can be expanded using the basis states as

$$
\begin{aligned}
O &= \left(\sum_n |\Psi_n\rangle \langle \Psi_n| \right) O \left(\sum_m |\Psi_m\rangle \langle \Psi_m| \right) \\
&= \sum_{nm} \langle \Psi_n| O |\Psi_m\rangle |\Psi_n\rangle \langle \Psi_m| \\
&= \sum_{nm} O_{nm} |\Psi_n\rangle \langle \Psi_m|, \quad (3.44)
\end{aligned}
$$

where

$$
\boxed{O_{nm} = \langle \Psi_n| O |\Psi_m\rangle} \quad (3.45)
$$

is the matrix element of O in this representation. The expectation value of operator O in this representation is then given by

$$
\begin{aligned}
\langle O \rangle &= \langle \Psi| O |\Psi\rangle = \sum_{nm} \psi_n^* \psi_m \langle \Psi_n| O |\Psi_m\rangle \\
&= \sum_{nm} \psi_n^* O_{nm} \psi_m \quad (3.46)
\end{aligned}
$$

If operator O is hermitian, then its matrix representation O_{nm} is a hermitian matrix

$$
O_{mn}^* = \langle \Psi_m| O |\Psi_n\rangle^* = \langle \Psi_n| O^\dagger |\Psi_m\rangle = \langle \Psi_n| O |\Psi_m\rangle^* = O_{nm}. \quad (3.47)
$$

3.1.6 REPRESENTATION OF SCHRÖDINGER EQUATION

If $H = (H_{nm})$ is the matrix of the Hamiltonian represented in a complete basis set, then the matrix representation of the Schrödinger equation is given by

$$
i\hbar \frac{\partial}{\partial t} \psi_n = \sum_m H_{nm} \psi_m. \quad (3.48)
$$

It can be also written as a matrix equation

$$
i\hbar \frac{\partial}{\partial t} \Psi = H\Psi, \quad (3.49)
$$

where Ψ is the wave function vector in this representation.

3.1.7 FEYNMAN-HELLMANN THEOREM

If the Hamiltonian H is a differential function of parameter λ, and E_n is an eigen-energy with $|n\rangle$ the corresponding normalized eigenstate,

$$
H|n\rangle = E_n|n\rangle, \quad (3.50)
$$

then it can be shown that

$$
\boxed{\frac{\partial E_n}{\partial \lambda} = \left\langle n \left| \frac{\partial H}{\partial \lambda} \right| n \right\rangle.} \quad (3.51)
$$

This is known as the Feynman-Hellmann theorem. It can be also written as

$$\boxed{\frac{\partial}{\partial \lambda} \langle n|H|n \rangle = \left\langle n \left| \frac{\partial H}{\partial \lambda} \right| n \right\rangle .}$$

(3.52)

It indicates that the derivative of the expectation value of the Hamiltonian equals the expectation value of the Hamiltonian's derivative in an energy eigenstate.

To prove this theorem, let us take the derivative for Eq. (3.50) with respect to λ. This leads to

$$\frac{\partial H}{\partial \lambda}|n\rangle + H \frac{\partial}{\partial \lambda}|n\rangle = \frac{\partial E_n}{\partial \lambda}|n\rangle + E_n \frac{\partial}{\partial \lambda}|n\rangle.$$

(3.53)

Multiplying both sides with $\langle n|$, and using the property $\langle n|H = \langle n|E_n$, we then obtain Eq. (3.51).

3.2 BASIS TRANSFORMATION

Supposing that $\{|\Psi_n\rangle\}$ and $\{|\Phi_n\rangle\}$ are two sets of complete orthonormal basis sets, the wave function of $|\Psi\rangle$ in these two basis sets are related by

$$\begin{aligned} \langle \Psi_n|\Psi\rangle &= \langle \Psi_n| \left(\sum_m |\Phi_m\rangle \langle \Phi_m| \right) |\Psi\rangle \\ &= \sum_m \langle \Psi_n|\Phi_m\rangle \langle \Phi_m|\Psi\rangle = \sum_m U_{nm} \langle \Phi_m|\Psi\rangle, \end{aligned}$$

(3.54)

where

$$\boxed{U_{nm} = \langle \Psi_n|\Phi_m\rangle}$$

(3.55)

defines a unitary transformation matrix between $\{|\Psi_n\rangle\}$ and $\{|\Phi_n\rangle\}$. U is unitary because

$$\begin{aligned} \left(UU^\dagger\right)_{nn'} &= \sum_m U_{nm} U^*_{n'm} \\ &= \sum_m \langle \Psi_n|\Phi_m\rangle \langle \Phi_m|\Psi_{n'}\rangle \\ &= \langle \Psi_n|\Psi_{n'}\rangle \\ &= \delta_{nn'}, \end{aligned}$$

(3.56)

and similarly

$$\left(U^\dagger U\right)_{mm'} = \delta_{mm'}.$$

(3.57)

The matrix representations of an operator A in these two sets of basis states, i.e. $A_{nm} = \langle \Psi_n|O|\Psi_m\rangle$ and $A'_{ij} = \langle \Phi_i|O|\Phi_j\rangle$, are also related by this unitary matrix

$$\begin{aligned} A_{nm} &= \sum_{ij} \langle \Psi_n|\Phi_i\rangle \langle \Phi_i|O|\Phi_j\rangle \langle \Phi_j|\Psi_m\rangle \\ &= \sum_{ij} U_{ni} A'_{ij} \left(U^\dagger\right)_{jm}, \end{aligned}$$

(3.58)

or simply

$$A = UA'U^\dagger. \tag{3.59}$$

Thus if A' is Hermitian, so is A

$$A^\dagger = UA'^\dagger U^\dagger = UA'U^\dagger = A. \tag{3.60}$$

Suppose ψ_n is an eigenvector of A with the eigenvalue a_n, namely

$$A\psi_n = a_n\psi_n, \tag{3.61}$$

then using Eq. (3.59), we obtain

$$A'\left(U^\dagger\psi_n\right) = a_n\left(U^\dagger\psi_n\right). \tag{3.62}$$

Therefore, a_n is also an eigenvalue of A'. The corresponding eigenvection is $U^\dagger\psi_n$. This means that the eigen-spectrum of an operator is not changed by unitary transformations.

3.2.1 EXAMPLE: FROM REAL TO MOMENTUM SPACE REPRESENTATION

As an example, let us consider the transformation between the coordinate eigenstates and the momentum eigenstates in one dimension. In this case, $|\Psi_n\rangle = |x\rangle$ and $|\Phi_n\rangle = |p\rangle$, and the transformation of a quantum state $|\Psi\rangle$ between these two basis sets is

$$\Psi(x) = \langle x|\Psi\rangle = \int dp\langle x|p\rangle\langle p|\Psi\rangle = \frac{1}{\sqrt{2\pi}}\int dp e^{ipx/\hbar}\Psi(p), \tag{3.63}$$

which is nothing but the Fourier transformation of $\Psi(p)$. The inverse transformation is

$$\Psi(p) = \langle p|\Psi\rangle = \int dx\langle p|x\rangle\langle x|\Psi\rangle = \frac{1}{\sqrt{2\pi}}\int dx e^{-ipx/\hbar}\Psi(x). \tag{3.64}$$

To find the representation of the coordinate operator $\hat{x}$ (here a 'hat' is added to x to emphasize that it is an operator) in momentum space, let us use $\hat{x}$ to act on an arbitrary quantum space $|\Psi\rangle$. In the momentum representation,

$$\begin{aligned} \langle p|\hat{x}|\Psi\rangle &= \int dxdp'x\langle p|x\rangle\langle x|p'\rangle\langle p'|\Psi\rangle \\ &= i\hbar\frac{\partial}{\partial p}\int dxdp'\langle p|x\rangle\langle x|p'\rangle\Psi(p') \\ &= i\hbar\frac{\partial}{\partial p}\int dp'\delta(p-p')\Psi(p') \\ &= i\hbar\frac{\partial}{\partial p}\Psi(p). \end{aligned} \tag{3.65}$$

Therefore, we have

$$\hat{x} = i\hbar\frac{\partial}{\partial p}. \tag{3.66}$$

Similarly, in momentum space, the one-dimensional Schrödinger eigen-equation becomes

$$E\Psi(p) = \left[\frac{p^2}{2\mu} + V\left(i\hbar\frac{\partial}{\partial p}\right)\right]\Psi(p). \tag{3.67}$$

3.3 COMMUTATORS

In classical mechanics, any two physical variables are commutable (or exchangeable). However, this property is no longer valid for physical operators. Commutation of operators is an important property of quantum mechanics.

The commutator of two operators A and B is defined by

$$\boxed{[A,B] = AB - BA.} \tag{3.68}$$

If the commutator of A and B vanishes, these two operators are said to commute.

3.3.1 PROPERTIES OF COMMUTABLE OPERATORS

Theorem: Two commutable operators A and B, i.e. $AB = BA$, possess a complete set of common eigenstates.

For example, in a one-dimensional translation invariant system, the momentum operator p_x commutes with the Hamiltonian H. These two operators have the common eigenstates, which are just the plane-wave states

$$\Psi(x) = \frac{1}{\sqrt{2\pi}}e^{ipx/\hbar}. \tag{3.69}$$

The corresponding eigenvalues of p_x and H are p and $p^2/2\mu$, respectively. The eigenstates of H are doubly degenerate for each given energy.

We prove the above theorem in two steps:

1. If $|n\rangle$ is a nondegenerate eigenstate of A with the eigenvalue a_n, then

$$A\left(B|n\rangle\right) = B\left(A|n\rangle\right) = a_n\left(B|n\rangle\right). \tag{3.70}$$

Thus $B|n\rangle$ is also an eigenstate of A with the same eigenvalue a_n. Since this eigenvalue is nondegenerate, $B|n\rangle$ can differ from $|n\rangle$ only by a multiplicative constant which is denoted as b_n

$$B|n\rangle = b_n|n\rangle. \tag{3.71}$$

Therefore, $|n\rangle$ is simultaneously an eigenfunctin of the operators A and B with the eigenvalues a_n and b_n, respectively.

2. If a_n is a α-fold degenerate eigenvalue of A and $|n\rangle = (|n,1\rangle, \cdots |n,\alpha\rangle)$ are the corresponding linearly independent eigenstates. Following the previous step, it is simple to show that $B|n,r\rangle$ is also an eigenstate of A with

the eigenvalue a_n. Thus $B|n, r\rangle$ can be expanded in terms of the α linearly independent eigenstates of A

$$B|n, r\rangle = \sum_{s=1}^{\alpha} C_{sr}|n, s\rangle, \tag{3.72}$$

where C_{rs} are the expansion coefficients, which forms an $\alpha \times \alpha$ square matrix. This equation can be also written as

$$B|n\rangle = |n\rangle C. \tag{3.73}$$

By diagonalizing C with a canonical matrix V

$$C = V\Lambda V^{-1}, \tag{3.74}$$

where Λ is a diagonal matrix. Substituting it into Eq. (3.73), we find that

$$B(|n\rangle V) = (|n\rangle V)\Lambda. \tag{3.75}$$

The column vectors of $(|n\rangle V)$ are therefore the eigenstates of B and the diagonal matrix elements of Λ are the corresponding eigenvalues. This shows that $(|n\rangle V)$ are simultaneously the eigenstates of A and B.

On the other hand, if two operators, A and B, do not commute, generally they do not have the same eigenstates.

3.3.2 PROPERTIES OF NONCOMMUTABLE OPERATORS

I. Conjugate pair

An operators, A, is said to be canonically conjugate to another operator, B, if their commutator

$$\boxed{[A, B] = i\hbar.} \tag{3.76}$$

A and B are said to form a canonically conjugate pair. For example, the position operator is canonically conjugate to its corresponding momentum operator

$$\boxed{[x, p_x] = [y, p_y] = [z, p_z] = i\hbar.} \tag{3.77}$$

II. Trace of operators

Finite dimensional operators behave similarly as matrices. For example, if both A and B are finite dimensional operators, the trace of AB equals that of BA

$$\text{Tr}(AB) = \text{Tr}(BA). \tag{3.78}$$

This is a direct consequence of the matrix representations of these operators.

However, if both A and B are infinite dimensional operators and contain continuous spectra, Eq. (3.78) may not be always valid. For example,

$$\boxed{\text{Tr}(xp_x) \neq \text{Tr}(p_x x),} \tag{3.79}$$

otherwise

$$0 = \text{Tr}(xp_x - p_x x) = \text{Tr}(i\hbar), \tag{3.80}$$

which is clearly wrong.

III. Exponent of noncommutable operators

For any two classical variables, A and B, the exponent of their sum can be always factorized as a product of the exponent for each variable

$$e^{A+B} = e^A e^B. \tag{3.81}$$

For operators, this kind of factorization is generally not valid except if A and B satisfy some specific properties:

1. If A and B commute with each other, then the exponential function of their sum can be factorized as a product of two exponents

$$e^{A+B} = e^A e^B. \tag{3.82}$$

In this case, A and B behave just like classical variables.

2. Baker-Hausdorff-Campbell theorem: If A and B satisfy the following commutation relations

$$[A, [A, B]] = [B, [A, B]] = 0, \tag{3.83}$$

then

$$e^{A+B} = e^A e^B e^{-[A,B]/2} = e^B e^A e^{[A,B]/2}. \tag{3.84}$$

Example: If $A = x$ and $B = p_x$, then $[x, [x, p_x]] = [p_x, [x, p_x]]$ and

$$e^{x+p_x} = e^x e^{p_x} e^{-i\hbar/2} = e^{p_x} e^x e^{i\hbar/2} \tag{3.85}$$

3. Trotter-Suzuki decomposition: If ε is a small parameter, then for any two operators A and B, the following approximate expressions hold

$$e^{\varepsilon(A+B)} \approx e^{\varepsilon A} e^{\varepsilon B} + O(\varepsilon^2), \tag{3.86}$$

$$e^{\varepsilon(A+B)} \approx e^{\varepsilon A/2} e^{\varepsilon B} e^{\varepsilon A/2} + O(\varepsilon^3), \tag{3.87}$$

Eqs. (3.86) and (3.87) are the second- and third-order Trotter-Suzuki decomposition formulas, respectively. These decomposition formulas are frequently used in quantum Monte Carlo or other numerical simulations.

3.4 SCHRÖDINGER PICTURE

Although there are many representations of wave functions and observables connected by unitary transformations, it is useful to distinguish certain classes of representations, called pictures, which differ in the way the time evolution of the system is treated.

The Schrödinger and Heisenberg pictures are two commonly used ones. They are related by a unitary transformation. All physical quantities, such as eigenvalues and expectation values of observables are identical, independent on the picture chosen.

The Schrödinger picture is one in which the operators are time-independent but the wave function is time dependent. The time evolution of the wave function $\Psi(t)$ is determined by the Schrödinger equation

$$i\hbar \frac{\partial}{\partial t}\Psi(t) = H\Psi(t). \tag{3.88}$$

Since this equation is a first order differential equation in time, $\Psi(t)$ is completely determined once it is specified at an initial time $t = 0$. A formal solution of this time-dependent Schrödinger equation is given by

$$\Psi(t) = e^{-iHt/\hbar}\Psi(0). \tag{3.89}$$

The evolution of the expectation value of an observable O with time is given by

$$\begin{aligned}
\frac{d}{dt}\langle O \rangle &= \frac{d}{dt}\langle \Psi | O | \Psi \rangle \\
&= \langle \partial_t \Psi | O | \Psi \rangle + \langle \Psi | \partial_t O | \Psi \rangle + \langle \Psi | O | \partial_t \Psi \rangle \\
&= \frac{1}{i\hbar}\langle \Psi | [O,H] | \Psi \rangle + \langle \Psi | \partial_t O | \Psi \rangle,
\end{aligned} \tag{3.90}$$

so that

$$\boxed{\frac{d}{dt}\langle O \rangle = \frac{1}{i\hbar}\langle [O,H] \rangle + \langle \partial_t O \rangle.} \tag{3.91}$$

If O does not depend explicitly on time, i.e. $\partial_t O = 0$, this equation reduces to

$$\frac{d}{dt}\langle O \rangle = \frac{1}{i\hbar}\langle [O,H] \rangle. \tag{3.92}$$

Furthermore, if O commutes with H, this equation means that its expectation value does not vary in time. We call this observable a constant of the motion. A particular case is the Hamiltonian itself. For a time-independent Hamiltonian, $\partial H/\partial t = 0$, we have

$$\frac{d}{dt}\langle H \rangle = \frac{1}{i\hbar}\langle [H,H] \rangle = 0. \tag{3.93}$$

This is just the mathematical formulation of the energy conservation.

3.4.1 VIRIAL THEOREM

The virial theorem relates the kinetic energy to its potential in an energy eigenstate. In Eq. (3.92), if we set $O = \mathbf{r} \cdot \mathbf{p}$, we have

$$
\begin{aligned}
\frac{d}{dt} \langle \mathbf{r} \cdot \mathbf{p} \rangle &= \frac{1}{i\hbar} \langle [\mathbf{r} \cdot \mathbf{p}, H] \rangle \\
&= \frac{1}{i\hbar} \left\langle \left[\mathbf{r} \cdot \mathbf{p}, \frac{\mathbf{p}^2}{2\mu} + V(\mathbf{r}) \right] \right\rangle \\
&= \left\langle \frac{\mathbf{p}^2}{\mu} - \mathbf{r} \cdot [\nabla V(\mathbf{r})] \right\rangle \\
&= \langle 2T - \mathbf{r} \cdot [\nabla V(\mathbf{r})] \rangle,
\end{aligned} \tag{3.94}
$$

where T is the kinetic energy. If the particle is in one of the eigenstate of the Hamiltonian (not necessary to be in the ground state), then

$$
\frac{d}{dt} \langle \mathbf{r} \cdot \mathbf{p} \rangle = 0, \tag{3.95}
$$

so that

$$
\boxed{2 \langle T \rangle = \langle \mathbf{r} \cdot [\nabla V(\mathbf{r})] \rangle.} \tag{3.96}
$$

This equation is known as the virial theorem. It holds in arbitrary dimensions.

If the potential is of the form

$$
V(\mathbf{r}) = \alpha r^n \tag{3.97}
$$

with α a $\mathbf{r}$-independent constant, the virial theorem can be also expressed as

$$
\boxed{\langle T \rangle = \frac{n}{2} \langle V(\mathbf{r}) \rangle.} \tag{3.98}
$$

In obtaining the virial theorem, we have implicitly assumed that the wave function and the expectation values of the kinetic and potential energies are all square integrable. It implies that this theorem does not apply to a particle that is not in a bound state.

If $n = 2$ (i.e. harmonic potential, which will be discussed in the next chapter), the virial theorem says that $\langle T \rangle = \langle V \rangle$. If $n = -2$, on the other hand, $\langle T \rangle = -\langle V \rangle$ and the total energy $\langle T + V \rangle = 0$, which implies that the eigenenergy of the "bound state", if exists, is exactly zero.

3.4.2 EHRENFEST THEOREM

Ehrenfest theorem relates the time derivative of the expectation values of the position and momentum operators to the expectation value of the "force". It is obtained by applying Eq. (3.92) to the position and momentum operators

$$
\mu \frac{d}{dt} \langle \mathbf{r} \rangle = \langle \mathbf{p} \rangle, \tag{3.99}
$$

$$
\frac{d}{dt} \langle \mathbf{p} \rangle = -\langle \nabla V(\mathbf{r}) \rangle. \tag{3.100}
$$

Substituting Eq. (3.99) into Eq. (3.100), we obtain

$$\mu \frac{d^2}{dt^2} \langle \mathbf{r} \rangle = -\langle \nabla V(\mathbf{r}) \rangle. \tag{3.101}$$

At first glance, this equation looks like the Newtow's classical equation of motion. However, there is a subtle difference. If the time evolution of $\langle \mathbf{r} \rangle$ does satisfy Newton's second law, the "force" on the right side of the equation would have to be $-\nabla V(\langle \mathbf{r} \rangle)$, which is typically not the same as $-\langle \nabla V(\mathbf{r}) \rangle$.

For example, for a cubic potential $V(x) \propto x^3$ in one dimension, its derivative is proportional to x^2. In the case of classical motion, the Newton's force should be proportional to $\langle x \rangle^2$. However, in quantum mechanics, this "force" is proportional to $\langle x^2 \rangle \neq \langle x \rangle^2$.

$\langle \nabla V(\mathbf{r}) \rangle = \nabla V(\langle \mathbf{r} \rangle)$ holds when $V(\mathbf{r}) \propto r^2$ is quadratic and the "force" varies linearly with $\mathbf{r}$. This is the case of harmonic oscillator. The Ehrenfeld theorem implies that the expected position should follow the classical trajectory governed by Newton's second law.

3.5 HEISENBERG PICTURE

The Heisenberg picture is obtained from the Schrödinger picture by gauging the Schrödinger wave function $\Psi(t)$ with the unitary evolution operator $\exp(iHt/\hbar)$. This defines the Heisenberg wave function Ψ_H

$$\Psi_H = e^{iHt/\hbar} \Psi(t) = \Psi(0), \tag{3.102}$$

which is time-independent and coincides the Schrödinger wave function at $t = 0$. However, operators become time dependent

$$O_H(t) = e^{iHt/\hbar} O e^{-iHt/\hbar}. \tag{3.103}$$

From the time evolution of O_H

$$\frac{d}{dt} O_H(t) = \frac{1}{i\hbar} e^{iHt/\hbar} (OH - HO) e^{-iHt/\hbar} + e^{iHt/\hbar} \partial_t O e^{-iHt/\hbar}, \tag{3.104}$$

we find the Heisenberg equation of motion for the operator O_H

$$\frac{d}{dt} O_H(t) = \frac{1}{i\hbar} [O_H, H_H] + (\partial_t O)_H. \tag{3.105}$$

A particular example is the x-component of the position and momentum of a particle, x and p. Their Heisenberg equations of motions are given by

$$\frac{d}{dt} x_H = \frac{1}{i\hbar} [x_H, H_H], \tag{3.106}$$

$$\frac{d}{dt} p_H = \frac{1}{i\hbar} [p_H, H_H]. \tag{3.107}$$

Using the representation

$$x_H = i\hbar \frac{\partial}{\partial p_H}, \tag{3.108}$$

$$p_H = -i\hbar \frac{\partial}{\partial x_H}, \tag{3.109}$$

we obtain

$$\frac{dx_H}{dt} = \frac{\partial H_H}{\partial p_H}, \tag{3.110}$$

$$\frac{dp_H}{dt} = -\frac{\partial H_H}{\partial x_H}. \tag{3.111}$$

These equations are formally identical to Hamilton's canonical equations in classical mechanics. In fact, the Ehrenfest theorem can be readily derived from these equations. Thus the Heisenberg picture corresponds to a formulation of quantum dynamics which is formally close to classical mechanics.

3.6 UNCERTAINTY PRINCIPLE

Let $\langle A \rangle$ be the expectation value of A in a given normalized state $|\Psi\rangle$. The uncertainty of A is defined to be the square root of the mean-square deviation of A about its expectation value

$$\Delta A = \left\langle (A - \langle A \rangle)^2 \right\rangle^{1/2} = \left(\langle A^2 \rangle - \langle A \rangle^2 \right)^{1/2}. \tag{3.112}$$

Similarly, we can define the expectation value of B and its uncertainty.

General Heisenberg uncertainty relation:
For two arbitrary hermitian operators, A and B, the following inequality holds

$$\Delta A \Delta B \geq \frac{1}{2} |\langle [A, B] \rangle|. \tag{3.113}$$

Proof: To prove the above expression, we introduce a new operator

$$C = A + i\eta B, \tag{3.114}$$

where η is a real constant. Both A and B are Hermitian, but C is not. Apparently, $CC^\dagger = \left(CC^\dagger \right)^\dagger$ is a real and nonnegative operator

$$\langle CC^\dagger \rangle = \langle \Psi | CC^\dagger | \Psi \rangle = \langle C^\dagger \Psi | C^\dagger \Psi \rangle \geq 0. \tag{3.115}$$

This means that

$$
\begin{aligned}
f(\eta) &= \langle (A+i\eta B)(A-i\eta B) \rangle \\
&= \langle A^2 \rangle + \eta^2 \langle B^2 \rangle - i\eta \langle [A,B] \rangle \\
&= \langle A^2 \rangle + \langle B^2 \rangle \left[\eta - \frac{i\langle [A,B] \rangle}{2\langle B^2 \rangle} \right]^2 + \frac{\langle [A,B] \rangle^2}{4\langle B^2 \rangle} \\
&\geq 0.
\end{aligned}
\tag{3.116}
$$

As

$$
[A,B]^\dagger = -[A,B],
\tag{3.117}
$$

the expectation value of $[A,B]$ is purely imaginary

$$
\langle [A,B] \rangle^* = -\langle [A,B] \rangle.
\tag{3.118}
$$

$f(\eta)$ has a minimum at

$$
\eta_0 = \frac{i\langle [A,B] \rangle}{2\langle B^2 \rangle},
\tag{3.119}
$$

and the value of $f(\eta)$ at the minimum is

$$
f(\eta_0) = \langle A^2 \rangle + \frac{\langle [A,B] \rangle^2}{4\langle B^2 \rangle}.
\tag{3.120}
$$

Since $f(\eta_0)$ is nonnegative, we have

$$
\langle A^2 \rangle \langle B^2 \rangle \geq -\frac{\langle [A,B] \rangle^2}{4}.
\tag{3.121}
$$

Substituting A by $A - \langle A \rangle$ and B by $B - \langle B \rangle$, and considering that

$$
[A - \langle A \rangle, B - \langle B \rangle] = [A,B],
\tag{3.122}
$$

we thus prove the inequality (3.113).

Applying the above inequality to two canonically conjugate operators, we immediately obtain

---

Heisenberg uncertainty principle:
If A and B are a pair of canonical conjugate operators, i.e. $[A,B] = i\hbar$, then

$$
\Delta A \Delta B \geq \frac{\hbar}{2}.
\tag{3.123}
$$

By further applying it to the canonical conjugate pairs (x, p_x), (y, p_y), (z, p_z), we have

$$
\begin{aligned}
\Delta x \Delta p_x &\geq \frac{\hbar}{2}, \\[1em]
\Delta y \Delta p_y &\geq \frac{\hbar}{2}, \\[1em]
\Delta z \Delta p_z &\geq \frac{\hbar}{2}.
\end{aligned}
\tag{3.124}
$$

The Heisenberg uncertainty principle does not place any restriction on the precision with which a position measurement of a particle can be made. However, once such a measurement has been made and the particle is known to be confined to some region of extent Δx, the ensemble of such particle is subsequently described by a wave function also of extent Δx. Subsequent measurements of the momentum on each of the identical systems composing the ensemble will then produce a range of values with a spread of order $\hbar/(2\Delta x)$.

Before the position was measured, if the particle was in a state of definite momentum, the act of measuring the position would force the system into a state in which the momentum is no longer known exactly but has become uncertain by an amount $\Delta p > \hbar/(2\Delta x)$. Thus the uncertainty principle does not refer to the period before the state was prepared, but only to the current situation.

The inherent limitations on measurement imposed by the uncertainty principle has nothing to do with the experimental errors that occur in actual measurements.

3.7 THE TIME-ENERGY UNCERTAINTY PRINCIPLE

In nonrelativistic quantum mechanics, time t is a special parameter. Unlike space coordinate $\mathbf{r}$, t is not a dynamic observable.

Eq. (3.113) does not apply to time and its "conjugate" variable, because the average and deviation of an operator is defined between two states on the equal time. The deviation of time is clearly not well defined.

However, in quantum mechanics,

$$
H = i\hbar \frac{\partial}{\partial t},
\tag{3.125}
$$

and

$$
\left[t, i\hbar \frac{\partial}{\partial t} \right] = -i\hbar,
\tag{3.126}
$$

these two equations imply that time and energy still form a pair of conjugate variables, and these quantities should satisfy a generalized uncertainty principle:

$$
\Delta t \Delta E \geq \frac{\hbar}{2}.
\tag{3.127}
$$

In this equation, ΔE is defined as the deviation of the energy

$$\Delta E = \langle H^2 \rangle - \langle H \rangle^2. \tag{3.128}$$

However, Δt is not the standard deviation of collection of time measured. Instead, it is the time that takes a measured quantity to change substantially. Thus the definition of Δt depends on the physical quantity that is measured.

Let us consider the measurement of an observable O in a physical state $|\Psi\rangle$. We assume O itself is time independent, $\partial_t O = 0$. From the uncertainty relation between O and the Hamiltonian H, given by Eq. (3.113), we have

$$\Delta O \cdot \Delta E \geq \frac{1}{2} |\langle [O, H] \rangle|. \tag{3.129}$$

Using Eq. (3.91), this equation can be also expressed as

$$\Delta O \cdot \Delta E \geq \frac{\hbar}{2} \left| \frac{d}{dt} \langle O \rangle \right|. \tag{3.130}$$

The time-energy uncertainty relation (3.127) is obtained if we define

$$\boxed{\Delta t = \frac{\Delta O}{\left| \dfrac{d}{dt} \langle O \rangle \right|}.} \tag{3.131}$$

Clearly, Δt represents the amount of time it takes the expectation value of O to change by one standard deviation. It depends on which observable is measured as well as on the rate it changes with time.

There are special cases in which $\Delta t \to \infty$:

1. The initial state is stationary, that is in an energy eigenstate. In this case $\Delta E = 0$ and $d\langle O \rangle / dt = 0$.
2. O is a conserving quantity and $\langle O \rangle$ does not change with time, $d\langle O \rangle / dt = 0$.

3.8 PROBLEMS

1. Operators O_i are defined as follows

$$
\begin{aligned}
O_1 \Psi(x) &= \Psi^2(x), \\
O_2 \Psi(x) &= x^2 \Psi(x), \\
O_3 \Psi(x) &= \frac{d^2}{dx^2} \Psi(x), \\
O_4 \Psi(x) &= \sin[\Psi(x)], \\
O_5 \Psi(x) &= \int_0^x \Psi(y) dy.
\end{aligned}
$$

Which of these operators are linear? Which are Hermitian?

2. A particle of mass μ moves in one dimension under the influence of a potential V(x). Suppose it is in an energy eigenstate

$$\Psi(x) = e^{-x^2/2}$$

with energy $E = \hbar^2/2\mu$.

a. Find the potential $V(x)$.
b. Find the probability of the particle whose momentum is between p and $p + \delta p$.

3. P is the one-dimensional spatial reflection operator

$$P\Psi(x) = \Psi(-x).$$

Prove that P is a hermitian operator.

4. If $f(x)$ can be expanded in a polynomial in x, show

$$[f(x), p_x] = i\hbar \frac{df}{dx}.$$

5. $|\Psi_1\rangle$ and $|\Psi_2\rangle$ are two eigenstates of the hermitian operator A. Find the conditions under which $|\Psi_1\rangle + 2|\Psi_2\rangle$ is also an eigenstate of A.

6. A and B are two Hermitian operators and $A^2 = B^2 = 4$.

a. Find the eigenvalues of each operator.
b. If α is an eigenvalue of A with $|\alpha\rangle$ the corresponding eigenstate, and

$$\langle \alpha | B | \alpha \rangle = 0,$$

calculate the uncertainty of B, i.e ΔB, in the state $|\alpha\rangle$.
c. Using the uncertainty principle

$$\Delta A \Delta B \geq \frac{1}{2} |\langle [A, B] \rangle|$$

to deduce the value of $\langle \alpha | [A, B] | \alpha \rangle$.

7. Assume that a charmed quark of mass 1.5 GeV/c^2 is confined to a volume with linear dimension of the order of 1 fm, and its average momentum is comparable with the minimum uncertainty in its momentum. Show that the confined quark may be treated as a nonrelativistic particle, and estimate its average kinetic energy.

8. If both A and B commute with $[A,B]$, prove

$$[A, F(B)] = [A,B]F'(B),$$
$$e^A e^B = e^{A+B} e^{\frac{1}{2}[A,B]}.$$

where $F(B)$ is an operator function of B that can be expressed by the Taylor series expansion.

9. Prove the identity

$$e^{xA} B e^{-xA} = B + x[A,B] + \frac{x^2}{2}[A,[A,B]] + \frac{x^3}{3!}[A,[A,[A,B]]] + \cdots$$

10. An electron moving in a uniform electric field, $\mathbf{E} = \alpha \hat{x}$, is described by the Hamiltonian

$$H = -\frac{\hbar^2}{2\mu}\nabla^2 - \alpha x.$$

Find $d\langle \mathbf{p}\rangle/dt$.

11. Using the Heisenberg uncertainty principle, estimate the ground state energy of a particle in the following one-dimensional potentials:

a. an infinite square well potential

$$V(x) = \begin{cases} 0, & |x| < L/2, \\ +\infty, & |x| > L/2. \end{cases}$$

b. a harmonic potential

$$V(x) = \frac{1}{2}\mu\omega^2 x^2,$$

where μ is the mass of the particle and ω is the angular frequency of the harmonic oscillator.

12. Sum rule: Evaluate the commutator

$$[p,[p,H]],$$

and prove the following sum rule:

$$\sum_n (E_n - E_m)|\langle n|p|m\rangle|^2 = \frac{\hbar^2}{2}\left\langle m\left|\frac{\partial^2 V}{\partial x^2}\right|m\right\rangle,$$

where p is the momentum operator, $|n\rangle$ and $|m\rangle$ are eigenstates of the one-dimensional Hamiltonian

$$H = \frac{p^2}{2\mu} + V(x).$$

13. Sturm-Liouville equation: Suppose $\{|n\rangle\}$ is a time-independent and orthonormalized complete basis set and

$$|\Psi(t)\rangle = \sum_n \psi_n(t)|n\rangle$$

is the wave function of the time-dependent Schrödinger equation with the Hamiltonian $H(t)$. Show that the density matrix, which is defined by

$$\rho_{nm} = \psi_n(t)\psi_m^*(t),$$

satisfies the equation

$$i\hbar\frac{\partial\rho(t)}{\partial t} = [H(t),\rho(t)].$$

14. Prove the Ehrenfest theorem

$$\frac{d}{dt}\langle\mathbf{r}\rangle = \frac{\mathbf{p}}{\mu},$$

$$\frac{d}{dt}\langle\mathbf{p}\rangle = -\langle\nabla V\rangle.$$

15. An operator a and its hermitian conjugate $a^\dagger$ satisfy the following anti-commutation relations

$$\{a,a\} = \{a^\dagger,a^\dagger\} = 0, \qquad \{a,a^\dagger\} = 1,$$

where the anticommutator is defined by $\{A,B\} = AB + BA$. Find the eigenvalues of $a^\dagger a$.

4 Harmonic Oscillators

4.1 ONE-DIMENSIONAL HARMONIC OSCILLATOR

The quantum harmonic oscillator is a quantum-mechanical analog of the classical harmonic oscillator. Because an arbitrary potential can usually be approximated as a harmonic potential at the vicinity of a stable equilibrium point, it is one of the most important model systems in quantum mechanics. It is also one of the few quantum-mechanical systems for which an exact, analytical solution is known.

Let us first consider a one-dimensional harmonic oscillator, whose Hamiltonian is defined by

$$H = -\frac{\hbar^2}{2\mu}\frac{\partial^2}{\partial x^2} + \frac{1}{2}\mu\omega^2 x^2, \tag{4.1}$$

where ω is the angular frequency of the oscillator. We solve this Hamiltonian using a method first introduced by Dirac, which does not make reference to any particular representation.

4.1.1 LADDER OPERATORS

To solve the problem of harmonic oscillators, we introduce two operators

$$P = \frac{1}{\sqrt{\mu}}p = -\frac{i\hbar}{\sqrt{\mu}}\frac{\partial}{\partial x}, \tag{4.2}$$

$$X = \sqrt{\mu}\omega x, \tag{4.3}$$

where $p = -i\hbar\partial_x$ is the momentum operator. These two operators satisfy the commutation relation

$$[X, P] = i\hbar\omega, \tag{4.4}$$

which yields

$$[X + iP, X - iP] = 2\hbar\omega. \tag{4.5}$$

The above equation suggests us to introduce the operator

$$a = \frac{1}{\sqrt{2\hbar\omega}}(X + iP) = \frac{1}{\sqrt{2\hbar\mu\omega}}(\mu\omega x + ip). \tag{4.6}$$

From Eq. (4.5), it is simple to show that

$$\left[a, a^\dagger\right] = 1. \tag{4.7}$$

a and $a^\dagger$ are called the ladder operators.

DOI: 10.1201/9781003174882-4

a (similarly $a^\dagger$) is an odd parity operator, since it depends linearly on both x and p which are also odd parity operators. It changes sign under the spatial reflection transformation.

In terms of these latter operators, X and P can be written as

$$X = \sqrt{\frac{\hbar\omega}{2}}\left(a+a^\dagger\right),$$ (4.8)

$$P = -i\sqrt{\frac{\hbar\omega}{2}}\left(a-a^\dagger\right),$$ (4.9)

and

$$x = \sqrt{\frac{\hbar}{2\mu\omega}}\left(a+a^\dagger\right),$$ (4.10)

$$p = -i\sqrt{\frac{\mu\hbar\omega}{2}}\left(a-a^\dagger\right).$$ (4.11)

Substituting these expressions into Eq. (4.1), the Hamiltonian is found to be

$$H = \frac{1}{4}\hbar\omega\left[\left(a+a^\dagger\right)^2 - \left(a-a^\dagger\right)^2\right] = \hbar\omega\left(a^\dagger a + \frac{1}{2}\right).$$ (4.12)

4.1.2 EIGEN-SPECTRUM

From the commutation relation of the ladder operators, we have

$$\left[a^\dagger a, a\right] = -a,$$ (4.13)
$$\left[a^\dagger a, a^\dagger\right] = a^\dagger.$$ (4.14)

This suggests that if $|\alpha\rangle$ is an eigenstate of $a^\dagger a$ with the eigenvalue α, then $a|\alpha\rangle$ and $a^\dagger|\alpha\rangle$ are also the eigenstates of $a^\dagger a$. The corresponding eigenvalues equal $\alpha - 1$ and $\alpha + 1$, respectively

$$a^\dagger a(a|\alpha\rangle) = a\left(a^\dagger a|\alpha\rangle\right) - a|\alpha\rangle = (\alpha-1)a|\alpha\rangle,$$ (4.15)
$$a^\dagger a(a^\dagger|\alpha\rangle) = a^\dagger\left(aa^\dagger\right)|\alpha\rangle = (\alpha+1)a^\dagger|\alpha\rangle.$$ (4.16)

Thus $a^\dagger$ raises and a lowers the eigenvalues by 1. This is the reason why they are called the ladder operators. They are also called creation (or raising) and annihilation (or lowering) operators, respectively.

Since $a^\dagger a$ is a nonnegative operator, its eigenvalues should all be nonnegative. Let $|0\rangle$ be the lowest eigenstate of $a^\dagger a$, we must have

$$a|0\rangle = 0.$$ (4.17)

Otherwise, $a|0\rangle$ is also an eigenstate of $a^\dagger a$ with the eigenvalue 1 minus than that of $|0\rangle$, contrary to the hypothesis that $|0\rangle$ is the lowest eigenstate. Therefore, we have

$$a^\dagger a|0\rangle = 0, \tag{4.18}$$

namely the lowest eigenvalue of $a^\dagger a$ is 0.

From $|0\rangle$, we can construct all other eigenstates of $a^\dagger a$ by repeatedly operating with $a^\dagger$ on the state $|0\rangle$. The series of states such generated have the eigenvalues $1, 2, \cdots$. In general, the n'th eigenstate (after normalization) can be written as

$$|n\rangle = \frac{1}{\sqrt{n!}} \left(a^\dagger\right)^n |0\rangle, \tag{4.19}$$

and

$$a^\dagger a|n\rangle = n|n\rangle. \tag{4.20}$$

Thus the eigenvalue of the Hamiltonian is given by

$$E_n = \hbar\omega \left(n + \frac{1}{2}\right). \tag{4.21}$$

4.1.3 EIGENFUNCTION

If $\Psi_0(x)$ is the wave function of $|0\rangle$, namely

$$|0\rangle = \int dx \Psi_0(x) |x\rangle, \tag{4.22}$$

then from Eq. (4.17), we have

$$a|0\rangle = \frac{1}{\sqrt{2\hbar\mu\omega}} \int dx \left(\mu\omega x + \hbar\frac{d}{dx}\right) \Psi_0(x)|x\rangle = 0, \tag{4.23}$$

hence

$$\left(\mu\omega x + \hbar\frac{d}{dx}\right) \Psi_0(x) = 0. \tag{4.24}$$

Solving this equaiton, we find that

$$\Psi_0(x) = C_0 e^{-\alpha x^2}, \tag{4.25}$$

where $\alpha = \mu\omega/2\hbar$ and C_0 is the normalization constant. The ground state wave function is node free in the whole space.

Using Eq. (4.19), we can further find the wave functions for the other eigenests. Generally, the wave function of the n'th eigenstate is

$$\Psi_n(x) = \left(\mu\omega x - \hbar\frac{d}{dx}\right)^n e^{-\alpha x^2} \tag{4.26}$$

up to a normalization constant. The wave function of the first excitation state is

$$\Psi_1(x) = 2\mu\omega x e^{-\alpha x^2}. \tag{4.27}$$

It is an odd function of x, containing a node at $x = 0$.

In general, the n'th eigenfunction is parity even or odd under the spatial reflection if n is even or odd. The n's eigenfunction contains n nodes in the whole space.

4.1.4 OCCUPATION REPRESENTATION

If all the eigenstates are normalized, then

$$a|n\rangle = d_n|n-1\rangle, \tag{4.28}$$

where d_n is a normalization coefficient. Since both $|n\rangle$ and $|n-1\rangle$ are normalized, $\langle n|n\rangle = \langle n-1|n-1\rangle$, we must have

$$n = \langle n|a^\dagger a|n\rangle = d_n^2, \tag{4.29}$$

hence $d_n = \sqrt{n}$.

Therefore, the matrix element of a is

$$\boxed{\langle n-1|a|n\rangle = \sqrt{n}.} \tag{4.30}$$

Similarly, the matrix element of the adjoint operator $a^\dagger$ is

$$\boxed{\langle n+1|a^\dagger|n\rangle = \sqrt{n+1}.} \tag{4.31}$$

The eigenstates of $a^\dagger a$ present a convenient representation, which is called the occupation representation, to represent physical observables. In this representation, the Hamiltonian is diagonal

$$H = \hbar\omega \begin{pmatrix} 1/2 & 0 & 0 & \cdots \\ 0 & 3/2 & 0 & \cdots \\ 0 & 0 & 5/2 & \cdots \\ \vdots & \vdots & \vdots & \ddots \end{pmatrix}, \tag{4.32}$$

and a is a matrix whose only nonzero elements lie on the diagonal immediately above the main diagonal

$$a = \begin{pmatrix} 0 & 1 & 0 & 0 & \cdots \\ 0 & 0 & \sqrt{2} & 0 & \cdots \\ 0 & 0 & 0 & \sqrt{3} & \cdots \\ \vdots & \vdots & \vdots & \vdots & \ddots \end{pmatrix}. \tag{4.33}$$

4.2 COHERENT STATE

For the harmonic oscillator, a coherent state $|\alpha\rangle$ is an eigenstate of the lower operator a

$$\boxed{a\,|\alpha\rangle = \alpha\,|\alpha\rangle}$$

(4.34)

with α the corresponding eigenvalue.

The coherent state is not an eigenstate of the Hamiltonian. The average energy of the coherent state is

$$\bar{H} = \langle H \rangle = \langle \alpha | H | \alpha \rangle = \left(\alpha^2 + \frac{1}{2} \right) \hbar\omega.$$

(4.35)

The average of the Hamiltonian square is

$$\langle H^2 \rangle = \frac{1}{4}\langle \alpha | \left(2a^\dagger a + 1 \right)^2 (\hbar\omega)^2 | \alpha \rangle = \left(\alpha^4 + 2\alpha^2 + \frac{1}{4} \right) (\hbar\omega)^2.$$

(4.36)

Thus the energy derivation is

$$\Delta H = \sqrt{\langle \alpha | (H^2 - \bar{H}^2) | \alpha \rangle} = \alpha\hbar\omega.$$

(4.37)

4.2.1 MINIMUM UNCERTAINTY STATE

Coherent state is physically interesting because it is a minimum uncertainty state at which the Heisenberg uncertainty relation reaches its minimum.

$$\boxed{\Delta x \Delta p = \frac{\hbar}{2}.}$$

(4.38)

To prove this, let us first calculate the average values of the x and p, $\langle x \rangle = \langle \alpha | x | \alpha \rangle$ and $\langle p \rangle = \langle \alpha | p | \alpha \rangle$, using the ladder operators

$$\langle x \rangle = \sqrt{\frac{\hbar}{2\mu\omega}} \langle \alpha | a + a^\dagger | \alpha \rangle = \sqrt{\frac{\hbar}{2\mu\omega}} (\alpha + \alpha^*),$$

(4.39)

$$\langle p \rangle = -i\sqrt{\frac{\hbar\mu\omega}{2}} \langle \alpha | a - a^\dagger | \alpha \rangle = -i\sqrt{\frac{\hbar\mu\omega}{2}} (\alpha - \alpha^*).$$

(4.40)

Similarly, the expectation values of x^2 and p^2, $\langle x^2 \rangle = \langle \alpha | x^2 | \alpha \rangle$ and $\langle p^2 \rangle = \langle \alpha | p^2 | \alpha \rangle$, are found to be

$$\langle x^2 \rangle = \frac{\hbar}{2\mu\omega} \langle \alpha | (a + a^\dagger)^2 | \alpha \rangle = \frac{\hbar}{2\mu\omega} \left[1 + (\alpha + \alpha^*)^2 \right],$$

(4.41)

$$\langle p^2 \rangle = -\frac{\hbar\mu\omega}{2} \langle \alpha | (a - a^\dagger)^2 | \alpha \rangle = \frac{\hbar\mu\omega}{2} \left[1 - (\alpha - \alpha^*)^2 \right].$$

(4.42)

Their deviations are

$$\Delta x = \sqrt{\langle x^2 \rangle - \langle x \rangle^2} = \sqrt{\frac{\hbar}{2\mu\omega}}, \tag{4.43}$$

$$\Delta p = \sqrt{\langle p^2 \rangle - \langle p \rangle^2} = \sqrt{\frac{\hbar\mu\omega}{2}} \tag{4.44}$$

The product of Δx and Δp just yields Eq. (4.38).

4.2.2 WAVE FUNCTION OF THE COHERENT STATE

To find explicitly the wave function of the coherent state, we multiply Eq. (4.19) from the right hand side by $\langle\alpha|$ and then take their hermitian conjugate. This yields

$$\langle n|\alpha \rangle = \frac{1}{\sqrt{n!}} \langle 0| a^n |\alpha \rangle = \frac{\alpha^n}{\sqrt{n!}} \langle 0|\alpha \rangle. \tag{4.45}$$

We thus have

$$|\alpha \rangle = \langle 0|\alpha \rangle \sum_n \frac{\alpha^n}{\sqrt{n!}} |n \rangle. \tag{4.46}$$

$\langle 0|\alpha \rangle$ can be always set as real. It is determined by the normalization condition

$$\langle \alpha|\alpha \rangle = \langle 0|\alpha \rangle^2 \sum_n \frac{(\alpha\alpha^*)^2}{n!} = \langle 0|\alpha \rangle^2 e^{\alpha\alpha^*} = 1, \tag{4.47}$$

so that

$$\langle 0|\alpha \rangle = e^{-\alpha\alpha^*/2}. \tag{4.48}$$

Therefore, we have

$$\boxed{|\alpha \rangle = e^{-\alpha\alpha^*/2} \sum_n \frac{\alpha^n}{\sqrt{n!}} |n \rangle.} \tag{4.49}$$

It can be also expressed as

$$\boxed{|\alpha \rangle = e^{-\alpha\alpha^*/2} \sum_n \frac{(\alpha a^\dagger)^n}{n!} |0 \rangle = e^{\alpha a^\dagger - \alpha\alpha^*/2} |0 \rangle.} \tag{4.50}$$

The coherent states are not orthogonal to each other. The overlap between two coherent states is given by

$$\begin{aligned} \langle \beta|\alpha \rangle &= e^{-\alpha\alpha^*/2 - \beta\beta^*/2} \sum_{nm} \frac{(\beta^*)^n \alpha^m}{\sqrt{n!}\sqrt{m!}} \langle n|m \rangle \\ &= e^{-\alpha\alpha^*/2 - \beta\beta^*/2} \sum_n \frac{(\beta^*\alpha)^n}{n!} \\ &= e^{-\beta\beta^*/2 - \alpha\alpha^*/2 + \beta^*\alpha}. \end{aligned} \tag{4.51}$$

Nevertheless, $|\alpha\rangle$ still form a complete (in fact, an over-complete) basis set

$$\boxed{\int \frac{d^2\alpha}{\pi} |\alpha\rangle \langle\alpha| = \sum_{mn} \frac{1}{\pi \sqrt{m!n!}} |n\rangle \langle m| \int d^2\alpha e^{-\alpha\alpha^*} \alpha^{*m} \alpha^n = 1.} \tag{4.52}$$

The last equality holds because

$$\int d^2\alpha e^{-\alpha\alpha^*} \alpha^{*m} \alpha^n = \pi m! \delta_{mn} \tag{4.53}$$

and

$$\sum_n |n\rangle \langle n| = 1. \tag{4.54}$$

4.3 CHARGED PARTICLES IN AN ELECTROMAGNETIC FIELD

4.3.1 MINIMAL COUPLING

If a charged particle of charge e moves in an electromagnetic field, there is a Lorentz force acting on the particle:

$$\mathbf{F} = e\left(\mathbf{E} + \mathbf{v} \times \mathbf{B}\right) \tag{4.55}$$

The electric and magnetic field strengths, $\mathbf{E}$ and $\mathbf{B}$, can be expressed by the corresponding potentials, $\varphi(\mathbf{r},t)$ and $A(\mathbf{r},t)$, according to the formula

$$\mathbf{E} = -\nabla\varphi - \frac{\partial \mathbf{A}}{\partial t}, \tag{4.56}$$

$$\mathbf{B} = \nabla \times \mathbf{A}. \tag{4.57}$$

Here $\mathbf{A}$ is the vector potential and φ the Coulomb potential. Both $\mathbf{A}$ and φ are real. In classical mechanics, the motion of the particle is governed by the Hamiltonian function

$$\boxed{H = \frac{1}{2\mu}\left(\mathbf{p} - e\mathbf{A}\right)^2 + e\varphi.} \tag{4.58}$$

This is the simplest way to couple a charged particle with an applied electromagnetic field. It is to replace the momentum $\mathbf{p}$ by $\mathbf{p} - e\mathbf{A}$. The coupling resulted is called the minimal coupling and $\mathbf{p} - e\mathbf{A}$ is called canonical momentum. In electrodynamics, this minimal coupling is adequate to account for the interaction between a charged particle and electromagnetic fields.

In quantum mechanics, the minimal coupling is to replace $\mathbf{p}$ by the momentum operator, i.e. $\mathbf{p} = -i\hbar\nabla$. The corresponding Hamiltonian reads

$$\boxed{H = \frac{1}{2\mu}\left(-i\hbar\nabla - e\mathbf{A}\right)^2 + e\varphi.} \tag{4.59}$$

4.3.2 GAUGE INVARIANCE

The electric and magnetic are invariant under the following gauge transformation

$$
\boxed{
\begin{aligned}
\mathbf{A} &\rightarrow \mathbf{A} + \nabla \chi(\mathbf{r}, t), \\[2mm]
\varphi &\rightarrow \varphi - \frac{\partial \chi(\mathbf{r}, t)}{\partial t},
\end{aligned}
}
\tag{4.60}
$$

where χ is an arbitrary function. To ensure the Schrödinger equation

$$
i\hbar \frac{\partial}{\partial t} \Psi(\mathbf{r}, t) = H \Psi(\mathbf{r}, t)
\tag{4.61}
$$

to be invariant under this gauge transformation, the wave function must be changed according to the formula

$$
\boxed{
\Psi \rightarrow \exp\left(\frac{ie}{\hbar} \chi(\mathbf{r}, t) \right) \Psi.
}
\tag{4.62}
$$

This invariance of the Schrödinger equation suggests that $(\mathbf{p} - e\mathbf{A}, i\hbar \partial_t - e\varphi)$ can be regarded as a U(1) gauge invariant momentum and "energy" operators, respectively.

Eq. (4.59) can be also written as

$$
H = H_0 - \frac{e}{2\mu}(\mathbf{A} \cdot \mathbf{p} + \mathbf{p} \cdot \mathbf{A}) + \frac{e^2}{2\mu}\mathbf{A}^2 + e\varphi,
\tag{4.63}
$$

where H_0 is the Hamiltonian without a magnetic field. Thus the coupling of the particle with the magnetic filed is given by the product $\mathbf{A} \cdot \mathbf{p}$. The third term depends only on the vector gauge field.

If we take the Coulomb gauge, $\nabla \cdot \mathbf{A} = 0$, so that $\mathbf{p} \cdot \mathbf{A} = \mathbf{A} \cdot \mathbf{p}$, the above Hamiltonian becomes

$$
H = H_0 - \frac{e}{\mu}\mathbf{A} \cdot \mathbf{p} + \frac{e^2}{2\mu}\mathbf{A}^2 + e\varphi.
\tag{4.64}
$$

$\mathbf{A} \cdot \mathbf{p}$ is the minimal coupling that guarantees gauge invariance in quantum theory.

4.3.3 PROBABILITY CURRENT

In the presence of an external vector potential, the Hamiltonian is changed, so is the probability current operator. To determine the probability current operator, let us take the time derivative of probability density

$$
\frac{\partial}{\partial t}\rho(\mathbf{r}, t) = \left(\frac{\partial}{\partial t}\Psi^* \right)\Psi + \Psi^*\left(\frac{\partial}{\partial t}\Psi \right).
\tag{4.65}
$$

Using the Schrödinger equation (4.61) and its complex conjugate equation

$$
-i\hbar \frac{\partial}{\partial t} \Psi^*(\mathbf{r}, t) = H^* \Psi^*(\mathbf{r}, t),
\tag{4.66}
$$

where

$$H^* = H_0 + \frac{e}{2\mu}(\mathbf{A}\cdot\mathbf{p} + \mathbf{p}\cdot\mathbf{A}) + \frac{e^2}{2\mu}\mathbf{A}^2 + e\varphi, \tag{4.67}$$

we find that

$$
\begin{aligned}
\frac{\partial}{\partial t}\rho(\mathbf{r},t) &= \frac{1}{i\hbar}\left(\Psi^* H\Psi - \Psi H^*\Psi^*\right) \\
&= \frac{i\hbar}{2\mu}\left(\Psi^*\nabla^2\Psi - \Psi\nabla^2\Psi^*\right) + \frac{e}{2\mu}\Psi(\mathbf{A}\cdot\nabla + \nabla\cdot\mathbf{A})\Psi^* \\
&\quad + \frac{e}{2\mu}\Psi^*(\mathbf{A}\cdot\nabla + \nabla\cdot\mathbf{A})\Psi \\
&= \frac{i\hbar}{2\mu}\nabla\cdot(\Psi^*\nabla\Psi - \Psi\nabla\Psi^*) + \frac{e}{\mu c}\nabla\cdot(\Psi^*\Psi\mathbf{A}).
\end{aligned}
\tag{4.68}
$$

This leads to the equation of probability conservation

$$\boxed{\frac{\partial}{\partial t}\rho(\mathbf{r},t) + \nabla\cdot\mathbf{j}(\mathbf{r},t) = 0,} \tag{4.69}$$

where $\mathbf{j}(\mathbf{r},t)$ is the probability current density

$$\boxed{\mathbf{j}(\mathbf{r},t) = \frac{i\hbar}{2\mu}\left(\Psi\nabla\Psi^* - \Psi^*\nabla\Psi\right) - \frac{e}{\mu}|\Psi|^2\mathbf{A}.} \tag{4.70}$$

This probability current density can be also written as

$$\mathbf{j}(\mathbf{r},t) = \frac{1}{2\mu}\Psi[(-i\hbar\nabla - e\mathbf{A})\Psi]^* + \frac{1}{2\mu}\Psi^*(-i\hbar\nabla - e\mathbf{A})\Psi \tag{4.71}$$

so that its gauge invariance can be more easily identified.

4.3.4 AHARONOV-BOHM EFFECT

The gauge invariance of the Hamiltonian (4.59) seems to imply that the gauge potential $(\mathbf{A},\varphi)$ is not directly measurable and the effect of electromagnetic field $(\mathbf{E},\mathbf{B})$ is absent where $\mathbf{E}$ and $\mathbf{B}$ vanish. This is indeed true in classical electrodynamics. However, in 1959, Aharonov and Bohm showed that the vector potential can in fact affect the quantum behavior of a charged particle when it is moving through a region in which the vector gauge potential is finite but the magnetic field $\mathbf{B}$ itself is zero. This phenomenon is now called the Aharonov-Bohm effect.

To understand the Aharonov-Bohm effect, let us consider a charged particle constrained to move in a path above or below a solenoid that is placed perpendicular to the page and carries a steady current. The solenoid is assumed to be infinitely long so that the magnetic field B is uniform inside the solenoid. Outside the solenoid, the magnetic field outside vanishes $\mathbf{B} = \nabla \times \mathbf{A} = 0$, but the vector potential is finite

$$\mathbf{A}(\mathbf{r}) = \frac{\Phi}{2\pi r}\hat{\phi}, \tag{4.72}$$

where Φ is the magnetic flux through the solenoid. The field in the solenoid is unchanged, so the scalar potential $\varphi(\mathbf{r}) = 0$.

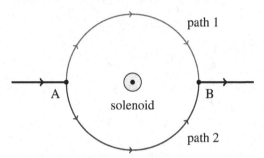

The curl-free gauge field $\mathbf{A}$ can in fact be removed out from the Hamiltonian (4.59) by taking a singular gauge transformation

$$\Psi(\mathbf{r},t) = e^{i\gamma(\mathbf{r})}\psi(\mathbf{r},t), \qquad (4.73)$$

where

$$\gamma(\mathbf{r}) = \frac{e}{\hbar} \int_{\mathbf{r}_0}^{\mathbf{r}} \mathbf{A}(\mathbf{r}) \cdot d\mathbf{r}, \qquad (4.74)$$

$\mathbf{r}_0$ is a reference point, and $d\mathbf{r}$ is a vector pointing along the path of the integral. As $\nabla \times \mathbf{A} = 0$, the line integral in this equation would not depend on the path from $\mathbf{r}_0$ to $\mathbf{r}$, provided that it does not wind the solenoid. In terms of ψ, the gradient of Ψ is

$$\nabla \Psi = e^{i\gamma} \left(\frac{ie}{\hbar}\mathbf{A} + \nabla \right) \psi, \qquad (4.75)$$

hence

$$(-i\hbar\nabla - e\mathbf{A})\Psi = -i\hbar e^{i\gamma}\nabla\psi. \qquad (4.76)$$

Substituting it into Eq. (4.61), and cancelling the common factor $\exp(i\gamma)$, we are left with

$$i\hbar\partial_t \psi(\mathbf{r},t) = -\frac{\hbar^2}{2\mu}\nabla^2 \psi(\mathbf{r},t). \qquad (4.77)$$

ψ satisfies the Schrödinger equation without the gauge potential $\mathbf{A}$. Therefore, the effect of the gauge potential is to add a phase factor to the wave function. The above gauge transformation is singular because the wave function such defined does not satisfy the periodic boundary condition if $\mathbf{r}$ runs around a closed path C that circulates the solenoid when the magnetic flux penetrating the solenoid is not an integer multiple of the magnetic flux quantum Φ_0

$$\Phi = \oint_C \mathbf{A} \cdot d\mathbf{r} \neq n\Phi_0, \qquad \Phi_0 = \frac{h}{e}, \qquad (4.78)$$

where n is an integer.

An important observation made by Aharonov and Bohm is that the phase gained for the particle propagating from A to B along the upper path is different from that along the lower path. The two paths form a closed loop C circulating the solenoid. The phase difference is

$$\Delta\gamma = \gamma_2 - \gamma_1 = \frac{e}{\hbar} \oint_C \mathbf{A}(\mathbf{r}) \times d\mathbf{r} = \frac{2\pi\Phi}{\Phi_0}, \tag{4.79}$$

where γ_1 are γ_2 are the phase factors gained from the upper and lower paths, respectively.

Apparently, the phase difference will cause an interference in the recombined wave function from the upper and lower paths:

$$\Psi_f \propto e^{i\gamma_1}\psi_f + e^{i\gamma_2}\psi_f = 2e^{i(\gamma_1+\gamma_2)}\psi_f \cos\frac{\Delta\gamma}{2}. \tag{4.80}$$

Therefore, the probability to observe a particle in the outgoing wave packet oscillates with the flux inside the solenoid

$$\langle\Psi_f|\Psi_f\rangle \propto \cos^2\frac{\Delta\gamma}{2} = \cos^2\frac{\pi\Phi}{\Phi_0}. \tag{4.81}$$

This quantum interference caused purely by a curl-free gauge potential was confirmed experimentally first by Robert G. Chambers[1] and later by many others.

4.4 LANDAU LEVELS

4.4.1 LANDAU GAUGE

Let us find the eigen-solution of a charged particle in a uniform magnetic field pointing along the z-axis, i.e. $\mathbf{B} = B\hat{z}$. We can fix the gauge by taking

$$\mathbf{A} = (-By, 0, 0), \tag{4.82}$$
$$\varphi = 0. \tag{4.83}$$

This is called the Landau gauge.

Under the above gauge, the Hamiltonian becomes

$$H = \frac{1}{2\mu}\left[(p_x + eBy)^2 + p_y^2 + p_z^2\right]. \tag{4.84}$$

It is simple to show that both p_x and p_z commute with the Hamiltonian. Thus the eigenvalues of both p_x and p_z are conserving quantities, and the eigenfunction can be decomposed as

$$\Psi(x,y,z) = e^{ik_x x + ik_z z}\psi(y). \tag{4.85}$$

---

[1]R. G. Chambers, Phys. Rev. Lett. **5**, 3 (1960).

Substituting it into the eigen-equation

$$H\Psi(x,y,z) = E\Psi(x,y,z), \tag{4.86}$$

we find that

$$\left[-\frac{\hbar^2}{2\mu}\frac{d^2}{dy^2} + \frac{1}{2\mu}\left(\hbar k_x + eBy\right)^2\right]\psi(y) = \varepsilon\psi(y), \tag{4.87}$$

where

$$\varepsilon = E - \frac{\hbar^2 k_z^2}{2\mu}. \tag{4.88}$$

To define

$$y' = y + \frac{\hbar k_x}{eB}, \tag{4.89}$$

$$\omega = \frac{eB}{\mu} \tag{4.90}$$

Eq. (4.87) is simplified to

$$\left(-\frac{\hbar^2}{2\mu}\frac{d^2}{dy'^2} + \frac{\mu\omega^2}{2}y'^2\right)\psi'(y') = \varepsilon\psi'(y'). \tag{4.91}$$

$\psi'(y') \equiv \psi(y' - \hbar k_x/eB)$ is just the eigenfunction of the harmonic oscillator centered at $y' = 0$. ω is the cyclotron frequency. From the discussion of Sec. 4.1, we know its eigenvalue is given by

$$\varepsilon = \hbar\omega\left(n + \frac{1}{2}\right), \tag{4.92}$$

hence

$$\boxed{E = \hbar\omega\left(n + \frac{1}{2}\right) + \frac{\hbar^2 k_z^2}{2\mu}.} \tag{4.93}$$

The first term represents the discrete energy eigenvalues corresponding to the motion in the plane orthogonal to the field, which are called Landau levels.

4.4.2 DEGENERACY OF LANDAU LEVELS

The Landau levels are highly degenerate because they do not depend on k_x. The degree of degeneracy of a Landau level is the number of possible values k_x. In a sample of finite width L_x, the allowed values of k_x are separated by $2\pi/L_x$.

In the Landau gauge, a Landau level is centered at

$$y_0 = \frac{\hbar k_x}{eB}. \tag{4.94}$$

The separation between two neighboring Landau levels is

$$\Delta y = \frac{h}{eBL_x}. \tag{4.95}$$

Thus the degeneracy of the Landau level is given by

$$\boxed{N = \frac{L_y}{\Delta y} = \frac{eBL_xL_y}{h} = \frac{\Phi}{\Phi_0},}$$ (4.96)

where

$$\Phi = BL_xL_y$$ (4.97)

is the total magnetic flux through the sample, and

$$\Phi_0 = \frac{h}{e}$$ (4.98)

is the flux quantum ($\Phi_0/2$ is the superconducting flux quantum). Thus, the degeneracy equals the number of flux quantum through the sample.

For a given (k_x, k_z), the above eigenfunction is localized in the y-direction, but not in the x-direction. This is due to the choice of Landau gauge. By taking a symmetric gauge, this in equivalence between x and y directions can be avoided.

4.4.3 SYMMETRIC GAUGE

Now let us consider the solution under the symmetric gauge

$$\mathbf{A} = \frac{1}{2}\mathbf{B} \times \mathbf{r} = \frac{B}{2}(-y, x, 0)$$ (4.99)

$$\varphi = 0.$$ (4.100)

Again the magnetic field $\mathbf{B}$ is applied along the z-axis. The Hamiltonian becomes

$$H = \frac{\mathbf{p}^2}{2\mu} - \frac{e}{2\mu}\mathbf{B} \cdot \mathbf{L} + \frac{e^2B^2(x^2+y^2)}{8\mu}.$$ (4.101)

The second term on the right hand side is just the Zeeman interaction between the orbital magnetic moment $\mathbf{m}_L = e\mathbf{L}/2\mu$ and the applied magnetic field. The above Hamiltonian can be also written as

$$H = \frac{1}{2\mu}\left(p_x + \frac{eBy}{2}\right)^2 + \frac{1}{2\mu}\left(p_y - \frac{eBx}{2}\right)^2 + \frac{p_z^2}{2\mu}.$$ (4.102)

To define

$$Q = \frac{1}{eB}\left(p_x + \frac{eBy}{2}\right),$$ (4.103)

$$P = p_y - \frac{eBx}{2},$$ (4.104)

we find that Q and P form a pair of conjugate operators that satisfy the canonical condition

$$[Q,P] = \frac{1}{eB}\left[p_x + \frac{eBy}{2}, p_y - \frac{eBx}{2}\right] = i\hbar.$$ (4.105)

Thus the Hamiltonian can be expressed as

$$H = \frac{P^2}{2\mu} + \frac{\mu}{2}\omega^2 Q^2 + \frac{p_z^2}{2\mu}. \tag{4.106}$$

Again p_z is conserved. The eigenfunction can be written as

$$\Psi(x,y,z) = e^{ik_z z}\psi(x,y) \tag{4.107}$$

and $\psi(x,y)$ is an eigenfunction of the first two terms in H, i.e.

$$H' = \frac{P^2}{2\mu} + \frac{\mu}{2}\omega^2 Q^2. \tag{4.108}$$

This is a harmonic oscillator. Its eigen-spectrum is given by

$$\varepsilon = \hbar\omega\left(n+\frac{1}{2}\right). \tag{4.109}$$

In terms of the ladder operator

$$a = \sqrt{\frac{\mu\omega}{2\hbar}}Q + i\sqrt{\frac{1}{2\hbar\mu\omega}}P \tag{4.110}$$

and its adjoint, we can write

$$H' = \hbar\omega\left(a^\dagger a + \frac{1}{2}\right). \tag{4.111}$$

The Landau level is highly degenerate because there is another pair of conjugate operators

$$Q' = p_y + \frac{eBx}{2}, \tag{4.112}$$

$$P' = \frac{1}{eB}\left(p_x - \frac{eBy}{2}\right), \tag{4.113}$$

which commutes with Q, P, and H'.

4.4.4 LOWEST LANDAU LEVEL

For the lowest Landau level, the eigenstate $\psi_{n=0}(x,y)$ satisfies the equation

$$a\psi_0(x,y) = 0, \tag{4.114}$$

which can be also expressed as

$$\left(\sqrt{\frac{\mu\omega}{2\hbar}}Q + i\sqrt{\frac{1}{2\hbar\mu\omega}}P\right)\psi_0(x,y) = 0 \tag{4.115}$$

or

$$\left[(p_x + ip_y) + \frac{eB}{2}(y - ix) \right] \psi_0(x,y) = 0. \tag{4.116}$$

If we introduce the complex coordinates

$$z = x + iy, \tag{4.117}$$
$$z^* = x - iy, \tag{4.118}$$

we can write Eq. (4.116) as

$$\left(\frac{\partial}{\partial z^*} + \frac{eB}{4\hbar} z \right) \psi_0(z, z^*) = 0. \tag{4.119}$$

Since the ground state of the standard Harmonic oscillator is a Gauss function of the coordinate, it is natural to assume that ψ_0 also contains a factor of Gauss function. This motivates us to set ψ_0 as

$$\psi_0(z, z^*) = \exp\left(-\frac{eB}{4\hbar} z z^* \right) u(z, z^*). \tag{4.120}$$

Inserting it into (4.119) and after simplification, we find that

$$\frac{\partial}{\partial z^*} u(z, z^*) = 0. \tag{4.121}$$

Thus for the lowest Landau level, $u(z, z^*)$ can be an arbitrary function of z. z^m ($m = 0, 1, ...$) serve as a linearly independent basis set. Thus the eigenfunction at the lowest Landau level can be a superposition of functions of the form

$$\boxed{\psi_{0,m}(z, z^*) = z^m \exp\left(-\frac{eB}{4\hbar} z z^* \right).} \tag{4.122}$$

The probability density of $\psi_{0,m}$ is

$$\rho_m(z, z^*)| = |z|^{2m} \exp\left(-\frac{|z|^2}{2r_0^2} \right), \tag{4.123}$$

where

$$r_0 = \sqrt{\frac{\hbar}{eB}} \tag{4.124}$$

is the magnetic length. From the derivative of $\rho_m|$ with respect to $|z|^2$, we find that ρ_m is roughly a Gaussian function concentrated at

$$|z_m| = \sqrt{2m} r_0. \tag{4.125}$$

The half-width of this Gaussian peak, Δz, is determined by the equation

$$\left(\frac{z_m + \Delta z}{z_m} \right)^{2m} \exp\left[m - m \left(\frac{z_m + \Delta z}{z_m} \right)^2 \right] = \frac{1}{2}. \tag{4.126}$$

In the large m limit, Δz is approximately found to be

$$\Delta z \approx \sqrt{\ln 2}\, r_0 \tag{4.127}$$

independent of m.

If the system is not infinite, but is a disc of radius R, then the largest value of m that can fit in is

$$M = \frac{\pi R^2 B}{\Phi_0}, \tag{4.128}$$

which is just the total number of magnetic flux penetrating the disc.

4.5 PROBLEMS

1. Calculate the position and momentum operators for the one-dimensional harmonic oscillator in the Heisenberg picture.

2. Half-harmonic oscillator: Find the exact eigenvalues for the potential

$$V(x) = \begin{cases} +\infty, & x < 0, \\ \dfrac{1}{2}\mu\omega^2 x^2, & x > 0. \end{cases}$$

3. Find the eigenvalues and the degeneracy of each eigenvalue for the three-dimensional harmonic oscillator defined by the Hamiltonian

$$H = -\frac{\hbar^2}{2\mu}\nabla^2 + \frac{1}{2}\mu\omega^2 r^2,$$

where $\mathbf{r} = (x, y, z)$ and ω is the angular frequency.

4. Hall current: an electron is confined to move in two dimensions, in an external electric field along the x-axis and an external magnetic field perpendicular to the two-dimensional plane. In the Landau gauge, the Hamiltonian is

$$H = \frac{1}{2\mu}\left[p_x^2 + (p_y - eBx)^2\right] - eEx.$$

a. Find the ground state eigenvalue and the corresponding eigenfunctions.
b. Find the expectation values of the canonical momentum operators along the x- and y-axis, $\langle p_x \rangle$ and $\langle p_y - eBx \rangle$, at each lowest-energy eigenstate.
c. Explain the results obtained in (b) using classical electrodynamics.

5. A particle in a one-dimensional harmonic potential is initially in the state

$$\Psi(x, t = 0) = e^{-\mu\omega(x-a)^2/2\hbar},$$

where μ is the mass and ω is the angular frequency.

a. Find $\Psi(x,t)$ and describe how it changes with time.
b. Calculate $\Delta x \cdot \Delta p$, p is the momentum operator.

6. Find the eigenvalues of an electron confined by a parabolic potential well along the x-direction

$$V(\mathbf{r}) = \frac{1}{2}\mu\omega^2 x^2$$

in the presence of a uniform and constant magnetic field B along the z-axis.

7. Find the ground state wave function in the momentum space representation of the one-dimensional harmonic oscillator

$$H = \frac{p^2}{2\mu} + \frac{1}{2}\mu\omega^2 x^2.$$

8. Lorentz force: the wave function of a particle of charge e moving through electric and magnetic fields $\mathbf{E}$ and $\mathbf{B}$ is governed by the Schrödinger equation

$$i\hbar\frac{\partial\Psi}{\partial t} = \left[\frac{1}{2\mu}(\mathbf{p}-e\mathbf{A})^2 + e\varphi\right]\Psi,$$

where the gauge potentials $\mathbf{A}$ and φ are related to the electric and magnetic fields by Eqs. (4.56–4.57). The magnetic field is assumed to be curl free

$$\nabla \times \mathbf{B} = 0.$$

a. Show that

$$\frac{d}{dt}\langle\mathbf{r}\rangle = \frac{1}{\mu}\langle\mathbf{p}-e\mathbf{A}\rangle.$$

b. If we identiry $d\langle\mathbf{r}\rangle/dt$ as the expectation value of the velocity operator $\langle\mathbf{v}\rangle$, and define the velocity operator $\mathbf{v}$ as

$$\mathbf{v} = \frac{1}{\mu}(\mathbf{p}-e\mathbf{A}),$$

show that

$$\mu\frac{d\langle\mathbf{v}\rangle}{dt} = e\mathbf{E} + e\langle\mathbf{v}\rangle\times\mathbf{B},$$

so the particle moves according to the Lorentz force law.

5 Angular Momentum

5.1 ORBITAL ANGULAR MOMENTUM

The angular momentum plays an important role in quantum mechanics. It is no longer an ordinary vector, instead it is a vector operator, whose three components do not commute with each other.

In classical mechanics, the orbital angular momentum of a particle is a vector $\mathbf{L}$ defined by the expression

$$\mathbf{L} = \mathbf{r} \times \mathbf{p}, \tag{5.1}$$

where $\mathbf{r}$ and $\mathbf{p}$ are the position and momentum vectors of the particle. This vector points in a direction at right angles to the plane containing $\mathbf{r}$ and $\mathbf{p}$.

In quantum mechanics, the angular momentum becomes a vector operator, which in the real-space representation is given by

$$\boxed{\mathbf{L} = -i\hbar \mathbf{r} \times \nabla,} \tag{5.2}$$

whose Cartesian components are the differential operators given by

$$L_x = -i\hbar \left(y\frac{\partial}{\partial z} - z\frac{\partial}{\partial y} \right), \tag{5.3}$$

$$L_y = -i\hbar \left(z\frac{\partial}{\partial x} - x\frac{\partial}{\partial z} \right), \tag{5.4}$$

$$L_z = -i\hbar \left(x\frac{\partial}{\partial y} - y\frac{\partial}{\partial x} \right). \tag{5.5}$$

Using the rules of commutators and the basic commutation relations between $\mathbf{r}$ and $\mathbf{p}$, we find the commutation relation satisfied by L_x, L_y, and L_z to be

$$[L_x, L_y] = i\hbar L_z, \tag{5.6}$$

$$[L_y, L_z] = i\hbar L_x, \tag{5.7}$$

$$[L_z, L_x] = i\hbar L_y. \tag{5.8}$$

These commutation relations can be also written as

$$\boxed{[L_\alpha, L_\beta] = i\hbar \varepsilon_{\alpha\beta\gamma} L_\gamma,} \tag{5.9}$$

where

$$\varepsilon_{\alpha\beta\gamma} = \begin{cases} 1 & \text{if } \alpha\beta\gamma = xyz \text{ or its rotations} \\ -1 & \text{if } \alpha\beta\gamma = yxz \text{ or its rotations} \\ 0 & \text{otherwise} \end{cases} \tag{5.10}$$

DOI: 10.1201/9781003174882-5

Thus any two components of the orbital angular momentum do not commute, and it is impossible to assign simultaneously definite values for all components of the angular momentum. It implies that if the system is in an eigenstate of one angular momentum component, it will generally not be in an eigenstate of either of the two other components.

The square of the orbital angular momentum is defined by

$$\mathbf{L}^2 = L_x^2 + L_y^2 + L_z^2. \tag{5.11}$$

Using the commutation relations of the angular momentum, it is simple to show that

$$\left[\mathbf{L}^2, L_\alpha\right] = 0, \quad (\alpha = x, y, z). \tag{5.12}$$

Thus one can find simultaneous eigenstates of $\mathbf{L}^2$ and one of the components of $\mathbf{L}$, so that both the magnitude of the orbital angular momentum and the value of one of its components can be determined precisely.

5.2 GENERAL ANGULAR MOMENTUM

A general angular momentum is defined as a vector operator $\mathbf{J}$ whose components are Hermitian operators satisfying the commutation relations similar to those for the orbital angular momentum

$$\boxed{\left[J_\alpha, J_\beta\right] = i\hbar\varepsilon_{\alpha\beta\gamma}J_\gamma.} \tag{5.13}$$

From this commutation relation, one can deduce that the square of the angular momentum

$$\mathbf{J}^2 = J_x^2 + J_y^2 + J_z^2 \tag{5.14}$$

commutes with J_x, J_y, and J_z, namely

$$\boxed{\left[\mathbf{J}^2, J_\alpha\right] = 0, \quad (\alpha = x, y, z).} \tag{5.15}$$

Now let us find the simultaneous eigenstates of $\mathbf{J}^2$ and J_z. The eigenvalues of $\mathbf{J}^2$ must be either positive or zero. For later convenience, we assume the eigenvalue of $\mathbf{J}^2$ to be $j(j+1)\hbar^2$ and that of J_z to be $m\hbar$. Denoting the simultaneous eigenstates of $\mathbf{J}^2$ and J_z by $|jm\rangle$, we have

$$\boxed{\begin{aligned} \mathbf{J}^2|jm\rangle &= j(j+1)\hbar^2|jm\rangle, \\ J_z|jm\rangle &= m\hbar|jm\rangle. \end{aligned}} \tag{5.16}$$

Since the expectation value of $\mathbf{J}^2$ is always larger than or equal to that of J_z^2

$$\langle\mathbf{J}^2\rangle = \langle J_x^2\rangle + \langle J_y^2\rangle + \langle J_z^2\rangle \geq \langle J_z^2\rangle, \tag{5.17}$$

we must have

$$j(j+1) \geq m^2. \tag{5.18}$$

We introduce the raising and lowering operators

$$\boxed{J_\pm = J_x \pm iJ_y.} \tag{5.19}$$

Apparently J_+ is adjoint to J_-, and vice versa

$$J_\pm^\dagger = J_\mp. \tag{5.20}$$

The commutation relations of between $J_\pm$ and $\left(\mathbf{J}^2, J_z\right)$ are found to be

$$
\begin{aligned}
\left[\mathbf{J}^2, J_\pm\right] &= 0, & (5.21)\\
\left[J_z, J_\pm\right] &= \pm\hbar J_\pm. & (5.22)\\
J_\pm J_\mp &= \mathbf{J}^2 - J_z^2 \pm \hbar J_z & (5.23)
\end{aligned}
$$

Using these relations, it is simple to show that $J_\pm\,|jm\rangle$ are also simultaneous eigenvectors of $\mathbf{J}^2$ and J_z, with the eigenvalues of $j(j+1)\hbar^2$ and $(m\pm1)\hbar$, respectively

$$
\begin{aligned}
\mathbf{J}^2\left(J_\pm\,|jm\rangle\right) &= J_\pm\mathbf{J}^2\,|jm\rangle = j(j+1)\hbar^2\left(J_\pm\,|jm\rangle\right), & (5.24)\\
J_z\left(J_\pm\,|jm\rangle\right) &= J_\pm J_z\,|jm\rangle \pm \hbar J_\pm\,|jm\rangle = (m\pm1)\hbar\left(J_\pm\,|jm\rangle\right). & (5.25)
\end{aligned}
$$

Thus by operating repeatedly with J_+, a sequence of eigenvectors of J_z can be generated, with eigenvalues $(m+1)\hbar$, $(m+2)\hbar$, and so on, each of which having the same eigenvalue of $\mathbf{J}^2$. Similarly, by operating repeatedly with J_-, a sequence of eigenvectors of J_z can be generated, with eigenvalues $(m-1)\hbar$, $(m-2)\hbar$, etc, and again each of which having the same eigenvalue of $\mathbf{J}^2$.

If $m_L\hbar$ is the largest eigenvalue of J_z, then

$$J_+\,|jm_L\rangle = 0. \tag{5.26}$$

Using Eq. (5.23), we obtain

$$
\begin{aligned}
J_- J_+\,|jm_L\rangle &= \left(\mathbf{J}^2 - J_z^2 - \hbar J_z\right)|jm_L\rangle \\
&= \left[j(j+1) - m_L^2 - m_L\right]\hbar^2\,|jm_L\rangle \\
&= 0. \tag{5.27}
\end{aligned}
$$

Thus

$$j(j+1) = m_L^2 + m_L. \tag{5.28}$$

Similarly, if $m_S\hbar$ is the smallest eigenvalue of J_z, then

$$J_-\,|jm_S\rangle = 0. \tag{5.29}$$

Using Eq. (5.23), we have

$$
\begin{aligned}
J_+ J_-\,|jm_S\rangle &= \left(\mathbf{J}^2 - J_z^2 + \hbar J_z\right)|jm_S\rangle \\
&= \left[j(j+1) - m_S(m_S-1)\right]\hbar^2\,|jm_S\rangle \\
&= 0. \tag{5.30}
\end{aligned}
$$

Hence

$$j(j+1) = m_S^2 - m_S. \tag{5.31}$$

The only solution that satisfies Eqs. (5.28) and (5.31) with $m_L \geq m_S$ is

$$m_L = -m_S = j \geq 0. \tag{5.32}$$

Moreover, $m_L - m_S = 2m_L$ must be an integer, therefore we have

$$\boxed{j = 0, \frac{1}{2}, 1, \frac{3}{2}, 2, \frac{5}{2}, \cdots} \tag{5.33}$$

and for a given j, the allowed values of m are

$$\boxed{m = -j, -j+1, \cdots, j-1, j.} \tag{5.34}$$

Thus the angular momentum j (and hence m) can take both integer and half-integer values.

5.2.1   MATRIX REPRESENTATION OF ANGULAR MOMENTUM OPERATORS

Here we consider the matrix representation of angular momentum operators in the space spanned by the orthonormal basis states of $\mathbf{J}^2$ and J_z

$$\langle j'm'|jm\rangle = \delta_{j'j}\delta_{m'm}. \tag{5.35}$$

Apparently, $\mathbf{J}^2$ and J_z are diagonal in this basis representation

$$\boxed{\begin{aligned} \langle j'm'|\mathbf{J}^2|jm\rangle &= j(j+1)\hbar^2\delta_{j'j}\delta_{m'm}, \\ \langle j'm'|J_z|jm\rangle &= m\hbar\delta_{j'j}\delta_{m'm}. \end{aligned}} \tag{5.36}$$

The matrix elements of $J_\pm$ can be also readily obtained using the equation

$$J_\pm|jm\rangle = c_\pm(j,m)\hbar|jm\pm 1\rangle, \tag{5.37}$$

where $c_\pm$ are normalization factors. Since both $|jm\rangle$ and $|jm\pm 1\rangle$ are normalized, we have

$$\begin{aligned} c_\pm^2(j,m)\hbar^2 &= \langle jm|J_\mp J_\pm|jm\rangle \\ &= \langle jm|\mathbf{J}^2 - J_z^2 \mp \hbar J_z|jm\rangle \\ &= [j(j+1) - m(m\pm 1)]\hbar^2, \end{aligned} \tag{5.38}$$

hence

$$c_\pm(j,m) = \sqrt{j(j+1) - m(m\pm 1)}. \tag{5.39}$$

Here we have implicitly assumed that $c_\pm$ is real and positive. Therefore, the matrix element of $J_\pm$ is given by

$$\boxed{\langle j'm'|J_\pm|jm\rangle = \hbar c_\pm(j,m)\delta_{j'j}\delta_{m'm\pm1}.} \tag{5.40}$$

The matrix representation of J_x and J_y can be found immediately from this expression using the relations

$$J_x = \frac{1}{2}(J_+ + J_-), \tag{5.41}$$

$$J_y = \frac{1}{2i}(J_+ - J_-). \tag{5.42}$$

For each given j, m can take $2j+1$ values, and J_x, J_y, J_z, and $\mathbf{J}^2$ are all $(2j+1)\times(2j+1)$ matrices. For $j=0$, the representation is trivial, J_x, J_y, J_z, and $\mathbf{J}^2$ are all one-dimensional zero matrix.

For $j=1/2$

$$J_x = \frac{\hbar}{2}\begin{pmatrix} 0 & 1 \\ 1 & 0 \end{pmatrix}, \tag{5.43}$$

$$J_y = \frac{\hbar}{2}\begin{pmatrix} 0 & -i \\ i & 0 \end{pmatrix}, \tag{5.44}$$

$$J_z = \frac{\hbar}{2}\begin{pmatrix} 1 & 0 \\ 0 & -1 \end{pmatrix}. \tag{5.45}$$

For $j=1$

$$J_x = \frac{\hbar}{\sqrt{2}}\begin{pmatrix} 0 & 1 & 0 \\ 1 & 0 & 1 \\ 0 & 1 & 0 \end{pmatrix}, \tag{5.46}$$

$$J_y = \frac{\hbar}{\sqrt{2}}\begin{pmatrix} 0 & -i & 0 \\ i & 0 & -i \\ 0 & i & 0 \end{pmatrix}, \tag{5.47}$$

$$J_z = \hbar\begin{pmatrix} 1 & 0 & 0 \\ 0 & 0 & 0 \\ 0 & 0 & -1 \end{pmatrix}. \tag{5.48}$$

5.3   EIGENFUNCTIONS OF ORBITAL ANGULAR MOMENTUM

We now find the eigenfunctions of the orbital angular momentum in the coordinate space.

In order to obtain the eigenvalues and eigenfunctions of $\mathbf{L}^2$ and L_z, it is more convenient to express the orbital angular momentum operators in spherical polar coordinates (r, θ, ϕ)

$$x = r \sin \theta \cos \phi, \tag{5.49}$$
$$y = r \sin \theta \sin \phi, \tag{5.50}$$
$$z = r \cos \theta, \tag{5.51}$$

with $0 \leq r < \infty, 0 \leq \theta \leq \pi$, and $0 \leq \phi \leq 2\pi$.

In the spherical polar coordinates, the orbital angular momentum becomes

$$
\begin{aligned}
L_x &= -i\hbar \left(-\sin\phi \frac{\partial}{\partial \theta} - \cot\theta \cos\phi \frac{\partial}{\partial \phi} \right), \\[2mm]
L_y &= -i\hbar \left(\cos\phi \frac{\partial}{\partial \theta} - \cot\theta \sin\phi \frac{\partial}{\partial \phi} \right), \\[2mm]
L_z &= -i\hbar \frac{\partial}{\partial \phi}.
\end{aligned}
\tag{5.52}
$$

and

$$
\mathbf{L}^2 = -\hbar^2 \left[\frac{1}{\sin\theta} \frac{\partial}{\partial \theta} \left(\sin\theta \frac{\partial}{\partial \theta} \right) + \frac{1}{\sin^2\theta} \frac{\partial^2}{\partial \phi^2} \right]. \tag{5.53}
$$

The corresponding lower or raising operators are

$$
L_+ = L_x + iL_y = \hbar e^{i\phi} \left(\frac{\partial}{\partial \theta} + i\cot\theta \frac{\partial}{\partial \phi} \right), \tag{5.54}
$$

$$
L_- = L_x - iL_y = -\hbar e^{-i\phi} \left(\frac{\partial}{\partial \theta} - i\cot\theta \frac{\partial}{\partial \phi} \right). \tag{5.55}
$$

$\mathbf{L}^2$ and L_z do not depend on r, their eigenstates are functions of (θ, ϕ) only. Assuming l and m are respectively the quantum numbers of $\mathbf{L}^2$ and L_z, and $Y_{lm}(\theta, \phi)$ the corresponding eigenfunction, we have

$$
\begin{aligned}
\mathbf{L}^2 Y_{lm}(\theta, \phi) &= l(l+1)\hbar^2 Y_{lm}(\theta, \phi), \\[2mm]
L_z Y_{lm}(\theta, \phi) &= m\hbar Y_{lm}(\theta, \phi).
\end{aligned}
\tag{5.56}
$$

$Y_{lm}(\theta, \phi)$ is just the so-called spherical harmonics. Since L_z depends on ϕ only, $Y_{lm}(\theta, \phi)$ can be factorized as a product of the eigenfunction of L_z and a function of θ

$$Y_{lm}(\theta, \phi) = \Phi_m(\phi)\Theta_{lm}(\theta), \tag{5.57}$$

where $\Phi_m(\phi)$

$$\Phi_m(\phi) = \frac{1}{\sqrt{2\pi}} \exp(im\phi) \tag{5.58}$$

is an eigenstate of L_z

$$L_z \Phi_m(\phi) = m\hbar \Phi_m(\phi). \tag{5.59}$$

From the periodicity of $\Phi_m(\phi)$

$$\Phi_m(\phi + 2\pi) = \Phi_m(\phi), \tag{5.60}$$

we know that m must be an integer. By comparison with the allowed values of m for the general angular momentum, this indicates that half-odd-integer values of j are excluded in the case of the orbital angular momentum.

For the eigenstate with $m = l$, we have

$$L_+ Y_{ll}(\theta, \phi) = \hbar e^{i\phi} \left(\frac{\partial}{\partial \theta} + i \cot \theta \frac{\partial}{\partial \phi} \right) \Phi_l(\phi) \Theta_{ll}(\theta) = 0. \tag{5.61}$$

It can be further simplified as

$$\left(\frac{\partial}{\partial \theta} - l \cot \theta \right) \Theta_{ll}(\theta) = 0. \tag{5.62}$$

The solution of this equation is given by

$$\Theta_{ll}(\theta) = C_l \sin^l \theta, \tag{5.63}$$

where C_l is a constant determined by the normalization condition,

$$\int_0^\pi d\theta \sin \theta |\Theta_{ll}(\theta)|^2 = 1. \tag{5.64}$$

Solving this equation, we find that

$$C_l = \frac{(-i)^l}{2^l l!} \sqrt{\frac{(2l+1)!}{2}}. \tag{5.65}$$

Here we have taken the convention to add a phase factor $(-i)^l$ to C_l so that the spherical harmonic function takes the standard form. Thus we have

$$\Theta_{ll}(\theta) = \frac{(-i)^l}{2^l l!} \sqrt{\frac{(2l+1)!}{2}} \sin^l \theta, \tag{5.66}$$

By applying L_- successively to $Y_{ll}(\theta, \phi)$, we can obtain all other eigenfunctions of $Y_{lm}(\theta, \phi)$. For example

$$
\begin{aligned}
Y_{l,l-1}(\theta, \phi) &= -\frac{1}{\sqrt{2l}\hbar} L_- Y_{l,l}(\theta, \phi) \\
&= -\frac{(-i)^l}{2^l l!} \sqrt{\frac{(2l+1)!}{8\pi l}} e^{i(l-1)\phi} \left(\frac{\partial}{\partial \theta} + l \cot \theta \right) \sin^l \theta \\
&= -\frac{(-i)^l}{2^l (l-1)!} \sqrt{\frac{(2l+1)!}{2\pi l}} e^{i(l-1)\phi} \cot \theta \sin^l \theta.
\end{aligned}
\tag{5.67}
$$

For the case $m = 0$,

$$\Theta_{l0}(\theta) = \sqrt{\frac{2l+1}{2}} P_l(\cos\theta), \tag{5.68}$$

where P_l is a Legendre polynomial.

In general,

$$Y_{lm}(\theta,\phi) = C_{lm} \frac{e^{im\phi}}{\sin^m\theta} \frac{d^{l-m}}{d(\cos\theta)^{l-m}} \sin^{2l}\theta, \tag{5.69}$$

where

$$C_{lm} = \frac{(-i)^l}{2^{l+1}l!} \sqrt{\frac{(2l+1)(l+m)!}{(l-m)!\pi}} \tag{5.70}$$

is the normalization constant. The expressions

$$\frac{d^l}{d(\cos\theta)^l} \sin^{2l}\theta \propto P_l(\cos\theta) \tag{5.71}$$

are the Legendre polynomials as function of $\sin\theta$ up to a constant factor.

For $l = m = 0$, we have

$$Y_{00}(\theta,\phi) = \frac{1}{\sqrt{4\pi}}. \tag{5.72}$$

States for $l = 0$ are called s-state (s stands for sharp). They are spherically symmetric.

For $l = 1$, we have

$$Y_{10}(\theta,\phi) = i\sqrt{\frac{3}{4\pi}} \cos\theta, \tag{5.73}$$

$$Y_{1\pm1}(\theta,\phi) = \mp i\sqrt{\frac{3}{8\pi}} e^{\pm i\phi} \sin\theta. \tag{5.74}$$

They are called p-states (stand for principal).

The wave functions of $l = 2$ are

$$Y_{20}(\theta,\phi) = -\sqrt{\frac{5}{16\pi}} (3\cos^2\theta - 1), \tag{5.75}$$

$$Y_{2\pm1}(\theta,\phi) = \pm\sqrt{\frac{15}{8\pi}} e^{\pm i\phi} \sin\theta\cos\theta, \tag{5.76}$$

$$Y_{2\pm2}(\theta,\phi) = -\sqrt{\frac{15}{32\pi}} e^{\pm 2i\phi} \sin^2\theta. \tag{5.77}$$

They are called d-states (stand for diffusive).

In the rectangular coordinates, the $l = 1$ states can be expressed as

$$Y_{10}(\theta,\phi) = i\sqrt{\frac{3}{4\pi}} \frac{z}{r}, \tag{5.78}$$

$$Y_{1\pm1}(\theta,\phi) = \mp i\sqrt{\frac{3}{8\pi}} \frac{x\pm iy}{r}. \tag{5.79}$$

Similarly, the $l = 2$ states are

$$Y_{20}(\theta, \phi) = -\sqrt{\frac{5}{16\pi}} \frac{3z^2 - r^2}{r^2}, \tag{5.80}$$

$$Y_{2\pm 1}(\theta, \phi) = \pm\sqrt{\frac{15}{8\pi}} \frac{(x \pm iy)z}{r^2}, \tag{5.81}$$

$$Y_{2\pm 2}(\theta, \phi) = -\sqrt{\frac{15}{32\pi}} \frac{(x \pm iy)^2}{r^2}. \tag{5.82}$$

5.4 SPIN ANGULAR MOMENTUM

Spin angular momentum is an intrinsic property of elementary particles. It is not associated with the motion of the particle in space, and is a quantum property without classical correspondence. To distinct from the orbital angular momentum, the spin angular momentum, or simply called spin, is generally denoted by **S**.

The spin operator **S** satisfies all the commutation relations of angular momentum. Electrons and nucleons (protons and neutrons), the building blocks of atomic and nuclear physics, have spin $1/2$. It is also believed that all the hadrons, including protons and neutrons, are made of more elementary constituents, the quarks, which also have spin $1/2$. Photons have spin 1.

5.4.1 PAULI MATRICES

For a particle of spin $S = 1/2$, m takes only two values, $m = \pm 1/2$. In the basis space where S_z is diagonal, the spin operator can be expressed as

$$S_\alpha = \frac{\hbar}{2}\sigma_\alpha \qquad \alpha = x, y, z, \tag{5.83}$$

where σ_α are the Pauli matrices defined by

$$\sigma_x = \begin{pmatrix} 0 & 1 \\ 1 & 0 \end{pmatrix}, \quad \sigma_y = \begin{pmatrix} 0 & -i \\ i & 0 \end{pmatrix}, \quad \sigma_z = \begin{pmatrix} 1 & 0 \\ 0 & -1 \end{pmatrix}. \tag{5.84}$$

The Pauli matrices satisfy the following commutation relations

$$[\sigma_x, \sigma_y] = 2i\sigma_z, \tag{5.85}$$
$$[\sigma_y, \sigma_z] = 2i\sigma_x, \tag{5.86}$$
$$[\sigma_z, \sigma_x] = 2i\sigma_y. \tag{5.87}$$

Moreover, $\sigma_x, \sigma_y, \sigma_z$ anticommute in pairs,

$$\sigma_\alpha \sigma_\beta = -\sigma_\beta \sigma_\alpha, \qquad \alpha \neq \beta. \tag{5.88}$$

They also satisfy the simple relation

$$\sigma_\alpha \sigma_\beta = i\sigma_\gamma, \tag{5.89}$$

where $\alpha, \beta, \gamma = x, y$ or z, in cyclic order. The traces of all Pauli matrices vanish.

$$Tr\sigma_\alpha = 0, \qquad \alpha = x, y, z, \tag{5.90}$$

and their determinants are equal to -1

$$\det \sigma_\alpha = -1, \qquad \alpha = x, y, z, \tag{5.91}$$

The square of σ_α $(\alpha = x, y, z)$ is an unity matrix

$$\sigma_\alpha^2 = I. \tag{5.92}$$

Using the above equations, it can be shown that

$$(\sigma \cdot \mathbf{A})(\sigma \cdot \mathbf{B}) = \mathbf{A} \cdot \mathbf{B} + i\sigma \cdot (\mathbf{A} \times \mathbf{B}), \tag{5.93}$$

where $\mathbf{A}$ and $\mathbf{B}$ are two vectors or two vector operators whose components commute with σ. A particular example is

$$(\sigma \cdot \mathbf{r})(\sigma \cdot \mathbf{p}) = \mathbf{r} \cdot \mathbf{p} + i\sigma \cdot (\mathbf{r} \times \mathbf{p}). \tag{5.94}$$

5.4.2 EIGENSTATES OF $S = 1/2$

Let $\mathbf{n}$ be a unit vector with polar angles (θ, ϕ). The component of the spin $\mathbf{S}$ along $\mathbf{n}$ is

$$S_n = \mathbf{n} \cdot \mathbf{S}. \tag{5.95}$$

S_n has two eigenvalues, $\pm\hbar/2$. We now find the eigenstates of S_n. The eigen-equation is given by

$$S_n \chi_\pm = \pm\frac{\hbar}{2}\chi_\pm, \tag{5.96}$$

where

$$\begin{aligned} S_n &= S_x \sin\theta \cos\phi + S_y \sin\theta \sin\phi + S_z \cos\theta \\ &= \frac{\hbar}{2}\left(\begin{array}{cc} \cos\theta & e^{-i\phi}\sin\theta \\ e^{i\phi}\sin\theta & -\cos\theta \end{array} \right). \end{aligned} \tag{5.97}$$

By solving Eq. (5.96), we find that when the eigenvalue of S_n is $\hbar/2$, the corresponding eigenvector is

$$\chi_+ = \left(\begin{array}{c} \cos\frac{\theta}{2} \\ e^{i\phi}\sin\frac{\theta}{2} \end{array} \right). \tag{5.98}$$

The expectation values of the three components of the spin vector are

$$\langle \mathbf{S} \rangle = \frac{\hbar}{2}\left(\sin\theta\cos\phi, \quad \sin\theta\sin\phi, \quad \cos\theta \right) = \frac{\hbar}{2}\mathbf{n}. \tag{5.99}$$

Thus the polar angles of $\mathbf{n}$ give the direction of $\langle \mathbf{S} \rangle$. In this sense we say the spin is up in the direction of $\mathbf{n}$.

When the eigenvalue of S_n is $-\hbar/2$, the eigenfunction is

$$\chi_- = \begin{pmatrix} \sin\frac{\theta}{2} \\ -e^{i\phi}\cos\frac{\theta}{2} \end{pmatrix}. \tag{5.100}$$

Upon calculating the expectation value of $\langle \mathbf{S} \rangle$, we find that

$$\langle \mathbf{S} \rangle = -\frac{\hbar}{2}\mathbf{n}. \tag{5.101}$$

Thus this is a "down"-spin state in the **n**-direction.

If $\mathbf{n} = (0,0,1)$ points along the z-axis, $\theta = 0$ and ϕ can take any value from 0 to 2π. For convenience, we take $\phi = \pi$, then

$$\chi_+ = \begin{pmatrix} 1 \\ 0 \end{pmatrix}, \quad \chi_- = \begin{pmatrix} 0 \\ 1 \end{pmatrix}. \tag{5.102}$$

These are just the two eigenstates of σ_z.

Similarly, one can also find the eigenfunctions of an arbitrary spin S, in an arbitrary direction **n** by directly solving the eigen-equation of S_n in its matrix representation.

5.4.3 QUBIT AND BLOCH SPHERE

In quantum computing, a qubit or quantum bit is a quantum version of the classical binary bit physically realized in a two-state (or two-level) quantum-mechanical system. This system can be realized, for example, by a $S = 1/2$ spin whose up and down spin states are taken as the two basis levels. In a classical system, a bit would have to be in one level or the other. However, in quantum mechanics, a qubit is allowed to be in a linear superposition of both levels simultaneously.

$$\Psi = a\begin{pmatrix} 1 \\ 0 \end{pmatrix} + b\begin{pmatrix} 0 \\ 1 \end{pmatrix}, \tag{5.103}$$

where a and b are two complex numbers defined on the four-dimensional unit sphere, $|a|^2 + |b|^2 = 1$, if Ψ is normalization. However, the states such defined are not all independent because any two states differing from each other by a global phase factor are physically equivalent. In other words, only the relative phase between the two basis states is physically relevant. It implies that we can always set the first coefficient a to be real and nonnegative, and parameterize a and b as

$$a = \cos\frac{\theta}{2}, \quad b = e^{i\phi}\sin\frac{\theta}{2}, \tag{5.104}$$

where $0 \le \theta \le \pi$ and $0 \le \phi < 2\pi$. Ψ then becomes

$$\Psi(\theta,\phi) = \cos\frac{\theta}{2}\begin{pmatrix} 1 \\ 0 \end{pmatrix} + e^{i\phi}\sin\frac{\theta}{2}\begin{pmatrix} 0 \\ 1 \end{pmatrix} = \begin{pmatrix} \cos\frac{\theta}{2} \\ e^{i\phi}\sin\frac{\theta}{2} \end{pmatrix}. \tag{5.105}$$

Parameters θ and ϕ can be interpreted as the spherical coordinates that specify a point

$$\mathbf{r} = (\sin\theta\cos\phi, \sin\theta\sin\phi, \cos\theta) \qquad (5.106)$$

on the surface of the three-dimensional unit sphere. This unit sphere is called Bloch sphere, named after the physicist Felix Bloch.

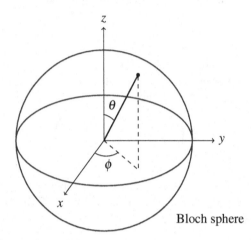

Bloch sphere

The Bloch sphere yields a geometrical representation of qubit—each point on the surface represents a qubit state. It is easy to read off all possible bases from this representation. In particular, orthogonal states correspond to antipodal points of the Bloch sphere because

$$\langle\Psi(\pi-\theta,\pi+\phi)|\Psi(\theta,\phi)\rangle = \cos\frac{\pi-\theta}{2}\cos\frac{\theta}{2} + e^{-i\pi}\sin\frac{\pi-\theta}{2}\sin\frac{\theta}{2} = 0. \quad (5.107)$$

For example, the north ($\theta = 0$) and south ($\theta = \pi$) poles correspond to the up and down spin states of σ_z, respectively. Similarly, the two antipodal points on the x-axis, i.e $(\theta,\phi) = (\pi/2,0)$ and $(\pi/2,\pi)$, correspond to the two eigenstates of σ_x

$$\Psi\left(\frac{\pi}{2},0\right) = \frac{1}{\sqrt{2}}\begin{pmatrix}1\\1\end{pmatrix}, \qquad \Psi\left(\frac{\pi}{2},\pi\right) = \frac{1}{\sqrt{2}}\begin{pmatrix}1\\-1\end{pmatrix}. \qquad (5.108)$$

Apparently, every pair of antipodal points corresponds to a basis set for the qubit.

5.5   ADDITION OF TWO ANGULAR MOMENTA

Here we consider the addition of two commuting angular momenta

$$\mathbf{J} = \mathbf{J}_1 + \mathbf{J}_2. \qquad (5.109)$$

where $\mathbf{J}_1$ and $\mathbf{J}_2$ are any two angular momenta. It is straightforward to show that $\mathbf{J}_1^2$ and $\mathbf{J}_2^2$ commute with $\mathbf{J}^2$ and J_z

$$\begin{aligned}\left[\mathbf{J}_1^2, J_z\right] &= \left[\mathbf{J}_1^2, \mathbf{J}^2\right] = 0, & (5.110)\\ \left[\mathbf{J}_2^2, J_z\right] &= \left[\mathbf{J}_2^2, \mathbf{J}^2\right] = 0. & (5.111)\end{aligned}$$

Thus the simultaneous eigenstates of $\mathbf{J}^2$ and J_z are also the eigenstates of $\mathbf{J}_1^2$ and $\mathbf{J}_2^2$. We denote these eigenstates as $|jm\rangle \equiv |j_1 j_2 jm\rangle$, where j is the eigenvalue of $\mathbf{J}^2$ and m is the eigenvalue J_z. These eigenstates can be represented using the basis functions of $\mathbf{J}_1^2$ and J_{1z}, $|j_1 m_1\rangle$, and those of $\mathbf{J}_2^2$ and J_{2z}, $|j_2 m_2\rangle$

$$|jm\rangle = \sum_{m_1 m_2} \langle j_1 m_1 j_2 m_2 | jm\rangle |j_1 m_1 j_2 m_2\rangle . \tag{5.112}$$

$\langle j_1 m_1 j_2 m_2 | jm\rangle \equiv \langle j_1 m_1 j_2 m_2 | j_1 j_2 jm\rangle$ are called Clebsch-Gordan coefficients. To solve the addition problem of two angular momenta is nothing but to determine the allowed values of j and m, and the rules governing the Clebsch-Gordan coefficients.

Since J_z commutes with both J_{1z} and J_{2z}, the eigenfuntion of J_{1z} and J_{2z} should also be an eigenfunction of J_z. It is simple to show

$$J_z |j_1 m_1 j_2 m_2\rangle = (m_1 + m_2)|j_1 m_1 j_2 m_2\rangle \tag{5.113}$$

thus we have

$$m = m_1 + m_2. \tag{5.114}$$

This means that the Clebsch-Gordan coefficients must vanish unless $m = m_1 + m_2$, and Eq. (5.112) can be rewritten as

$$|jm\rangle = \sum_{m_1 + m_2 = m} \langle j_1 m_1 j_2 m_2 | jm\rangle |j_1 m_1 j_2 m_2\rangle . \tag{5.115}$$

Now we determine the values of j by considering the basis states spanned by $|j_1, m_1\rangle$ and $|j_2, m_2\rangle$:

1. Because the maximum allowed values of m_1 and m_2 are given respectively by j_1 and j_2, the maximum possible value of m is $j_1 + j_2$. It follows that the maximum possible value of j is $j_{max} = j_1 + j_2$. For $j = m = j_1 + j_2$, there is only one term in the sum on the right hand side of Eq. (5.115), namely the term with $m_1 = j_1$ and $m_2 = j_2$.

m_1	m_2	m	allowed value of j
j_1	j_2	$j_1 + j_2$	$j_1 + j_2$

2. Next, let us consider the states with $m = j_1 + j_2 - 1$. In this case, there are two combinations for the values of m_1 and m_2, either $m_1 = j_1 - 1$ and $m_2 = j_2$ or $m_1 = j_1$ and $m_2 = j_2 - 1$. Thus, we have two linearly independent basis states $|j_1, j_1 - 1, j_2, j_2\rangle$ and $|j_1, j_1, j_2, j_2 - 1\rangle$. One of their linear combinations belongs to the set of eigenstates of $j = j_1 + j_2$. The other, which is orthogonal to the former one, must be an eigenstate belongs to the set of eigenstates for which the maximum value of m is $j_1 + j_2 - 1$, hence $j = j_1 + j_2 - 1$.

m_1	m_2	m	allowed value of j
$j_1 - 1$	j_2	$j_1 + j_2 - 1$	$j_1 + j_2$
j_1	$j_2 - 1$		$j_1 + j_2 - 1$

3. Proceeding further to states with $m = j_1 + j_2 - 2$, we see that three linearly independent states of this kind exist, corresponding to the values $j = j_1 + j_2$, $j = j_1 + j_2 - 1$, and $j = j_1 + j_2 - 2$, respectively.

m_1	m_2	m	allowed value of j
$j_1 - 2$	j_2		$j_1 + j_2$
$j_1 - 1$	$j_2 - 1$	$j_1 + j_2 - 2$	$j_1 + j_2 - 1$
j_1	$j_2 - 2$		$j_1 + j_2 - 2$

4. Repeating the above argument successively, we find that the minimum value of j equals $|j_1 - j_2|$ because when j reaches this minimum value all the combinations have been exhausted. This can be readily checked by noting that the total number of basis states in the $|j_1 j_2 Jm\rangle$ representation should be equal to the total number of basis states in the $|j_1 m_1 j_2 m_2\rangle$ representation, which equals $(2j_1 + 1)(2j_2 + 1)$. For each value of j, there are $2j + 1$ allowed values of m, thus the total number of $|j_1 j_2 jm\rangle$ basis states equals

$$\sum_{j=|j_1-j_2|}^{j_1+j_2} (2j+1) = (2j_1 + 1)(2j_2 + 1) \qquad (5.116)$$

as expected. Therefore, the allowed values of j are

$$\boxed{j = |j_1 - j_2|, \quad |j_1 - j_2| + 1, \quad \cdots, \quad j_1 + j_2.} \qquad (5.117)$$

5.5.1 CLEBSCH-GORDAN COEFFICIENTS

$\langle j_1 m_1 j_2 m_2 | jm \rangle$ can be determined by applying the raising and lowering operators $J_\pm = J_x \pm iJ_y$ to Eq. (5.115) to obtain a recursion relation between these coefficients. In order to remove the ambiguity, the relative phases between $|jm\rangle$ and $|j_1 m_1 j_2 m_2\rangle$ must be specified. A frequently used convention, from that of Condon and Shortley and of Wigner, is to assume that

1. the superposed angular momentum state $|jm\rangle$ satisfies Eq. (5.37) with the coefficients precisely given by $c_\pm(j,m)$ without other phase factor, i.e.

$$J_\pm |jm\rangle = \hbar c_\pm(j,m)|j,m\pm 1\rangle, \qquad (5.118)$$

2. $\langle j_1 j_1, j_2, j - j_1 | jj \rangle$ is real and positive.

With this convention, the Clebsch-Gordan coefficients are real, so that

$$\langle jm|j_1 m_1 j_2 m_2\rangle = \langle j_1 m_1 j_2 m_2 | jm \rangle. \qquad (5.119)$$

Applying $J_\pm = J_{1,\pm} + J_{2,\pm}$ to Eq. (5.115), we obtain

$$
\begin{aligned}
& c_\pm(j,m)|j,m\pm 1\rangle \\
= & \sum_{m_1+m_2=m} \langle j_1 m_1 j_2 m_2 | jm \rangle c_\pm(j_1, m_1)|j_1 m_1 \pm 1, j_2 m_2\rangle \\
& + \sum_{m_1+m_2=m} \langle j_1 m_1 j_2 m_2 | jm \rangle c_\pm(j_2, m_2)|j_1 m_1 j_2 m_2 \pm 1\rangle.
\end{aligned}
\qquad (5.120)
$$

This yields the recursion relations for determining Clebsch-Gordan coefficients

$$
\begin{aligned}
&c_{\pm}(j,m)\langle j_1 m_1, j_2, m_2|j, m\pm 1\rangle \\
&= c_{\mp}(j_1, m_1)\langle j_1, m_1 \mp 1, j_2 m_2|jm\rangle \\
&\quad + c_{\mp}(j_2, m_2)\langle j_1 m_1 j_2 m_2 \mp 1|jm\rangle.
\end{aligned}
\tag{5.121}
$$

In literature, the Clebsch-Gordan coefficients are also frequently represented by Wigner 3-j symbols, defined by

$$
\begin{pmatrix} j_1 & j_2 & j \\ m_1 & m_2 & m \end{pmatrix} = \frac{(-)^{j_1+j_2-m}}{\sqrt{2j+1}}\langle j_1 m_1 j_2 m_2|j, -m\rangle.
\tag{5.122}
$$

5.5.2   ADDITION OF TWO $S = 1/2$ SPINS

Let us consider the addition of two $S = 1/2$ spins, $\mathbf{S}_1$ and $\mathbf{S}_2$. The total spin

$$
\mathbf{S} = \mathbf{S}_1 + \mathbf{S}_2.
\tag{5.123}
$$

For each $S = 1/2$ spin, there are two basis states, which we denote as

$$
|\uparrow\rangle = \left|\frac{1}{2}, \frac{1}{2}\right\rangle \qquad |\downarrow\rangle = \left|\frac{1}{2}, -\frac{1}{2}\right\rangle.
\tag{5.124}
$$

They are the simultaneous eigenstates of $(\mathbf{S}_1^2, S_{1z})$ or $(\mathbf{S}_2^2, S_{2z})$. The total dimension of the Hilbert space is $2 \times 2 = 4$.

The allowed values of the total spin S are 1 and 0. $S = 1$ has three states, which is called spin triplet, and m takes three values, 1, 0, and -1. The $S = 0$ state is a spin singlet, and $m = 0$ is the only allowed value.

We first consider the $S = 1$ state. The state with $S = m = 1$ is simply given by

$$
|1,1\rangle = |\uparrow\rangle_1 |\uparrow\rangle_2 = |\uparrow\uparrow\rangle.
\tag{5.125}
$$

Then, by applying the lowering operator $S_- = S_{1-} + S_{2-}$ to this state and using Eq. (5.118), we obtain

$$
S_-|1,1\rangle = \hbar(|\uparrow\downarrow\rangle + |\downarrow\uparrow\rangle) = \hbar\sqrt{2}|1,0\rangle,
\tag{5.126}
$$

hence

$$
|1,0\rangle = \frac{1}{\sqrt{2}}(|\uparrow\downarrow\rangle + |\downarrow\uparrow\rangle).
\tag{5.127}
$$

By further applying S_- to $|1,0\rangle$ and using Eq. (5.118), we obtain

$$
S_-|1,0\rangle = \sqrt{2}\hbar|\downarrow\downarrow\rangle = \sqrt{2}\hbar|1,-1\rangle,
\tag{5.128}
$$

so that

$$
|1,-1\rangle = |\downarrow\downarrow\rangle.
\tag{5.129}
$$

The basis state of $S = m = 0$ is a linear superposition of the basis states $|\uparrow\downarrow\rangle$ and $|\downarrow\uparrow\rangle$ which are orthogonal to $|1,0\rangle$. This orthogonal basis state is readily to find, which is given by

$$|0,0\rangle = \frac{1}{\sqrt{2}} (|\uparrow\downarrow\rangle - |\downarrow\uparrow\rangle). \tag{5.130}$$

From Eqs. (5.125), (5.127), (5.129), and (5.130), we find that the spin triplet and singlet are symmetric and antisymmetric under exchange of the two $S = 1/2$ spins, respectively. As an example, the ground state of the hydrogen molecule, which contains two electrons, is a spin singlet state, while the first excited state is a spin triplet state.

5.6 WIGNER-ECKART THEOREM*

The Wigner-Eckart theorem, which is derived by physicists Eugene Wigner and Carl Eckart, is a theorem about the matrix elements of irreducible tensor operators of rank l, $T^{(l)}$, between two angular momentum eigenstates. $T^{(l)}$ is also called the spherical tensor operator. It is defined as a set of $2l + 1$ operators, $\{T_k^{(l)} : k = -l, -l+1, \cdots, l\}$, that fulfills the following commutation relations with the angular momentum $\mathbf{J}$

$$\left[J_z, T_k^{(l)} \right] = k\hbar T_k^{(l)}, \tag{5.131}$$

$$\left[J_\pm, T_k^{(l)} \right] = c_\pm(l,k)\hbar T_{k\pm1}^{(l)}. \tag{5.132}$$

The Wigner-Eckart theorem states that in the eigenstate representation of the angular momentum, $\mathbf{J}^2$ and J_z, where the basis vectors are given by $|\alpha jm\rangle$, the matrix elements of the tensor operator, $\left\langle \alpha' j'm' | T_k^{(l)} | \alpha jm \right\rangle$, depend on the projection quantum numbers, (m, m', k), only through the Clebsch-Gordan coefficient

$$\boxed{\langle \alpha' j'm' | T_k^{(l)} | \alpha jm \rangle = \langle j'm' | jmlk \rangle \langle \alpha' j' \| T^{(l)} \| \alpha j \rangle.} \tag{5.133}$$

Here α is the other quantum number of the system. $\langle \alpha' j' \| T^{(l)} \| \alpha j \rangle$ does not depend on the projection quantum numbers (m, m', k), and is referred to as the reduced matrix element. $\langle jmlk | j'm' \rangle$ is the Clebsch-Gordan coefficient.

The Wigner-Eckart theorem indicates that the tensor operators of rank l behave like adding a state with angular momentum l to an angular momentum basis state. This theorem provides an explicit form for the dependence of all matrix elements of irreducible tensors on the projection quantum numbers and a formal expression of the conservation laws of angular momentum. It allows one to determine quickly the selection rules for the matrix element that follow from rotational invariance. In addition, if matrix elements must be calculated, this theorem offers a way of significantly reducing the computational effort.

Rank 0 tensors are typically called scalars. Examples include pure numbers and the inner product of any two vectors, like $\mathbf{r}^2$, $\mathbf{p}^2$, and $\mathbf{J}^2$.

Rank-1 tensors are typically called vectors. If we define the angular momentum operator as

$$\mathbf{J} = (J_{-1}, J_0, J_1) = \left(\frac{J_x - iJ_y}{\sqrt{2}}, J_z, \frac{-J_x - iJ_y}{\sqrt{2}} \right), \tag{5.134}$$

it is straightforward to show that (J_{-1}, J_0, J_1) fulfill the defining commutation relations of a rank-1 tensor. Therefore, the angular momentum is a rank-1 tensor. Similarly, we can show that the momentum $\mathbf{p}$ and coordinate $\mathbf{r}$ are rank-1 tensors.

5.6.1 PROOF OF THE WIGNER-ECKART THEOREM

To prove the theorem, let us consider $T_k^{(l)} |\alpha jm\rangle$ and a superposition state of it with fixed α and j

$$|\alpha \bar{j} \bar{m}\rangle = \sum_{m,q} T_q^{(l)} |\alpha jm\rangle \langle jmlk|\bar{j}\bar{m}\rangle = \sum_{m+q=\bar{m}} T_q^{(l)} |\alpha jm\rangle \langle jmlk|\bar{j}\bar{m}\rangle. \tag{5.135}$$

The second equality holds because the Clebsch-Gordan coefficient vanishs unless $\bar{m} = m + k$.

By applying J_z to $T_k^{(l)} |\alpha jm\rangle$, we find that

$$\begin{aligned} J_z T_k^{(l)} |\alpha jm\rangle &= \left[J_z, T_k^{(l)} \right] |\alpha jm\rangle + T_k^{(l)} J_z |\alpha jm\rangle \\ &= (k+m) \hbar T_k^{(l)} |\alpha jm\rangle. \end{aligned} \tag{5.136}$$

It indicates that $T_k^{(l)} |\alpha jm\rangle$ is an eigenstate of J_z with the eigenvalue $(k+m)\hbar$. Together with Eq. (5.135) yields

$$\begin{aligned} J_z |\alpha \bar{j}\bar{m}\rangle &= \sum_{m,k} J_z T_k^{(l)} |\alpha jm\rangle \langle jmlk|\bar{j}\bar{m}\rangle \\ &= \sum_{m,k} (k+m) \hbar T_k^{(l)} |\alpha jm\rangle \langle jmlk|\bar{j}\bar{m}\rangle \\ &= \bar{m}\hbar |\alpha \bar{j}\bar{m}\rangle. \end{aligned} \tag{5.137}$$

Similarly, we have

$$\begin{aligned} J_\pm T_k^{(l)} |\alpha jm\rangle &= \left[J_\pm, T_k^{(l)} \right] |\alpha jm\rangle + T_k^{(l)} J_\pm |\alpha jm\rangle \\ &= c_\pm(l,k) T_{k\pm 1}^{(l)} |\alpha jm\rangle + c_\pm(j,m) T_k^{(l)} |\alpha j, m \pm 1\rangle, \end{aligned} \tag{5.138}$$

and

$$\begin{aligned} J_\pm |\alpha \bar{j}\bar{m}\rangle &= \sum_{m,k} J_\pm T_k^{(l)} |\alpha jm\rangle \langle jmlk|\bar{j}\bar{m}\rangle \\ &= \sum_{m,k} \left[c_\pm(l,k) T_{k\pm 1}^{(l)} |\alpha jm\rangle + c_\pm(j,m) T_k^{(l)} |\alpha j, m \pm 1\rangle \right] \\ &\quad \langle jmlk|\bar{j}\bar{m}\rangle. \end{aligned} \tag{5.139}$$

Alternatively, we may write it as

$$J_\pm |\alpha \bar{j} \bar{m}\rangle = \sum_{m,k} T_k^{(l)} |\alpha jm\rangle [c_\mp (j,m) \langle j, m \mp 1, lk | \bar{j} \bar{m}\rangle$$
$$+ c_\mp (l,k) \langle jml, k \mp 1 | \bar{j} \bar{m}\rangle] \tag{5.140}$$

To simplify it using Eq. (5.121), we obtain

$$J_\pm |\alpha \bar{j} \bar{m}\rangle = c_\pm (\bar{j}, \bar{m}) \sum_{m,k} T_k^{(l)} |\alpha jm\rangle \langle jmlk | \bar{j}, \bar{m} \pm 1\rangle$$
$$= c_\pm (\bar{j}, \bar{m}) |\alpha \bar{j}, \bar{m} \pm 1\rangle. \tag{5.141}$$

Eqs. (5.137) and (5.141) show that the states $|\alpha \bar{j} \bar{m}\rangle$ fulfill the angular momentum algebra and are the eigenstates of J^2 and J_z. It implies that the scalar product of two superposition states

$$\langle \alpha' j'm' | \alpha \bar{j} \bar{m}\rangle = \delta_{j',\bar{j}} \delta_{m',\bar{m}} \langle \alpha' \bar{j} \bar{m} | \alpha \bar{j} \bar{m}\rangle. \tag{5.142}$$

Furthermore, it can be shown that $\langle \alpha' \bar{j} \bar{m} | \alpha \bar{j} \bar{m}\rangle$ does not depend on $\bar{m}$

$$\langle \alpha' \bar{j} \bar{m} | \alpha \bar{j} \bar{m}\rangle = \frac{1}{c_\mp (\bar{j}, \bar{m})} \langle \alpha' \bar{j} \bar{m} | J_\pm | \alpha \bar{j}, \bar{m} \mp 1\rangle = \langle \alpha' \bar{j}, \bar{m} \mp 1 | \alpha \bar{j}, \bar{m} \mp 1\rangle. \tag{5.143}$$

This suggests that $\langle \alpha' \bar{j} \bar{m} | \alpha \bar{j} \bar{m}\rangle$ is a matrix element

$$\langle \alpha' \bar{j} \bar{m} | \alpha \bar{j} \bar{m}\rangle = \langle \alpha' \bar{j} \| \alpha \bar{j}\rangle, \tag{5.144}$$

that can be used to define the reduced matrix element

$$\langle \alpha' \bar{j} \| T^{(l)} \| \alpha j\rangle \equiv \langle \alpha' \bar{j} \| \alpha \bar{j}\rangle. \tag{5.145}$$

By means of the orthogonality of the Clebsch-Gordan coefficient, we can represent $T_k^{(l)} |\alpha jm\rangle$ as

$$T_k^{(l)} |\alpha jm\rangle = \sum_{\bar{j} \bar{m}} |\alpha \bar{j} \bar{m}\rangle \langle \bar{j} \bar{m} | jmlk\rangle. \tag{5.146}$$

Multiplication with $\langle \alpha' j'm' |$ yields

$$\langle \alpha' j'm' | T_k^{(l)} | \alpha jm\rangle = \sum_{\bar{j} \bar{m}} \langle \alpha' j'm' | \alpha \bar{j} \bar{m}\rangle \langle \bar{j} \bar{m} | jmlk\rangle$$
$$= \langle \alpha' j' \| \alpha j'\rangle \langle j'm' | jmlk\rangle$$
$$= \langle \alpha' j' \| T^{(l)} \| \alpha j\rangle \langle j'm' | jmlk\rangle. \tag{5.147}$$

This proves the Wigner-Eckart theorem.

5.7 PROBLEMS

1. A particle of spin $S = 1$ is in an orbital angular momentum $L = 2$ state. Find the energy levels and degeneracies associated with a spin–orbit interaction term of the form

$$H = \alpha \mathbf{L} \cdot \mathbf{S},$$

where α is a constant.

2. Suppose an electron is in a state described by the wave function

$$\Psi(\mathbf{r}) = \frac{1}{\sqrt{4\pi}} \left(e^{i\phi} \sin\theta + \cos\theta \right) R(r),$$

where

$$\int_0^\infty R^2(r) r^2 dr = 1,$$

and ϕ and θ are azimuth and polar angles, respectively.

   a. What are the possible results of a measurement of the z-component of the orbital angular momentum of the electron in this state?
   b. What is the probability of obtaining each of the above results?

3. If operator A commutes with the angular momentum operators

$$[A, J_\alpha] = 0, \qquad (\alpha = x,\, y,\, z),$$

show that the matrix element

$$\langle J, m | A | J, m' \rangle$$

is finite only if $m = m'$ and its value is independent on m.

4. Consider a beam of particles with $l = 1$. A measurement of L_x yields the result $\hbar$. What values will be obtained by a subsequent measurement of L_z, and with what probabilities?

5. For a system whose orbital angular momentum equals an odd integer, show that the three eigenstates, $|L_x = 0\rangle$, $|L_y = 0\rangle$ and $|L_z = 0\rangle$, are orthogonal to each other.

6. $\mathbf{a} = (a_x, a_y, a_z)$ is a vector with every component a two-dimensional hermitian operator. Its projection along any spatial direction $\mathbf{n}$, i.e. $\mathbf{n} \cdot \mathbf{a}$, takes the same two eigenvalues, ± 1. Find the conditions under which $\hbar a_x$, $\hbar a_y$, and $\hbar a_z$ satisfy the commutation relations of the angular momentum.

7. A deuterium atom is composed of a nucleus of spin 1 and an electron. The electronic angular momentum is a superposition of its orbital angular momentum and spin. If the orbital angular momentum of the electron is in either a s-wave ($L = 0$) or a p-wave ($L = 1$) state, what are the possible values of the total angular momentum of the deuterium atom.

8. An electron is in the $S_z = +\hbar/2$ state.

   a. If the x-component of the electron spin is measured, what is the probability of each possible measurement result?

    b. Same as for (a), but now the axis defining the measured spin direction makes an angle θ with respect to the z-axis.

9. $|jm\rangle$ is a common eigenstate of the angular momentum operators $\mathbf{J}^2$ and J_z, evaluate the expectation values of $J_x J_y$ and J_x^2 in this state.

6 Central potential

6.1 THREE-DIMENSIONAL POTENTIAL WITH SPHERICAL SYMMETRY

The Schrödinger eigen-equation in three dimensions reads

$$\left[-\frac{\hbar^2}{2\mu} \nabla^2 + V(\mathbf{r}) \right] \Psi(\mathbf{r}) = E\Psi(\mathbf{r}). \tag{6.1}$$

This is a second-order partial differential equation, in contrast to the one-dimensional case where the Schrödinger eigen-equation is a second-order ordinary differential equation. This equation can be solved by dimension reduction when the variables are separatable. A particularly interesting system is that of central potential, where the potential depends only on the magnitude $r = |\mathbf{r}|$ of the vector $\mathbf{r}$

$$V(\mathbf{r}) = V(r). \tag{6.2}$$

For such spherical potential, it is simple to solve the Schrödinger equation in the spherical polar coordinates, where ∇^2 is

$$\nabla^2 = \frac{1}{r^2} \frac{\partial}{\partial r} r^2 \frac{\partial}{\partial r} + \frac{1}{r^2} \left(\frac{1}{\sin\theta} \frac{\partial}{\partial \theta} \sin\theta \frac{\partial}{\partial \theta} + \frac{1}{\sin^2\theta} \frac{\partial^2}{\partial \phi^2} \right). \tag{6.3}$$

Using the expression of the orbital angular momentum operator, Eq. (5.53), it can be also expressed as

$$\nabla^2 = \frac{1}{r^2} \frac{\partial}{\partial r} r^2 \frac{\partial}{\partial r} - \frac{\mathbf{L}^2}{\hbar^2 r^2}. \tag{6.4}$$

In this spherical polar representation, the Hamiltonian is

$$H = -\frac{\hbar^2}{2\mu r^2} \frac{\partial}{\partial r} \left(r^2 \frac{\partial}{\partial r} \right) + \frac{\mathbf{L}^2}{2\mu r^2} + V(r). \tag{6.5}$$

It can be also expressed as

$$H = -\frac{\hbar^2}{2\mu r} \frac{\partial^2}{\partial r^2} r + \frac{\mathbf{L}^2}{2\mu r^2} + V(r), \tag{6.6}$$

where $\mathbf{L}^2$ is the square of the orbital angular momentum, defined in Eq. (5.53), which depends only on the polar angles (θ, ϕ) but not on r. The Schrödinger equation then becomes

$$\left[-\frac{\hbar^2}{2\mu r} \frac{\partial^2}{\partial r^2} r + \frac{\mathbf{L}^2}{2\mu r^2} + V(r) \right] \Psi(\mathbf{r}) = E\Psi(\mathbf{r}). \tag{6.7}$$

DOI: 10.1201/9781003174882-6

$\mathbf{L}^2$ does not depend on r. Clearly, it commutes with the Hamiltonian, and has a common eigenfunction with the Hamiltonian. If we use $Y_{lm}(\theta, \phi)$ to represent the eigenstate of $\mathbf{L}^2$ and L_z, with $l(l+1)\hbar^2$ the eigenvalue of $\mathbf{L}^2$ and $m\hbar$ the eigenvalue of L_z

$$\mathbf{L}^2 Y_{lm}(\theta, \phi) = l(l+1)\hbar^2 Y_{lm}(\theta, \phi), \tag{6.8}$$

$$L_z Y_{lm}(\theta, \phi) = m\hbar Y_{lm}(\theta, \phi), \tag{6.9}$$

then the wave function can be factorized as a product of a r-dependent function, $R_{nl}(r)$, and $Y_{lm}(\theta, \phi)$

$$\boxed{\Psi(\mathbf{r}) = R_{nl}(r)Y_{lm}(\theta, \phi) = \frac{1}{r}u_{nl}(r)Y_{lm}(\theta, \phi),} \tag{6.10}$$

where

$$u_{nl}(r) = rR_{nl}(r). \tag{6.11}$$

In $R_{nl}(r)$ or $u_{nl}(r)$, n is a quantum number which remains to be determined. $u_{nl}(r)$ does not depend on the magnetic quantum m because $\mathbf{L}^2$ does not depend on m.

Substituting Eq. (6.10) into Eq. (6.7), we obtain the differential equation for the radial function $u_{nl}(r)$

$$\left[-\frac{\hbar^2}{2\mu}\frac{d^2}{dr^2} + V_{\text{eff}}(r) \right] u_{nl}(r) = Eu_{nl}(r), \tag{6.12}$$

where

$$V_{\text{eff}}(r) = \frac{l(l+1)\hbar^2}{2\mu r^2} + V(r) \tag{6.13}$$

is an effective potential which, in addition to the interaction potential $V(r)$, also contains the repulsive centrifugal barrier term that is inversely proportional to r^2. It holds only for positive r. The radial function $R_{nl}(r)$ should be finite at the origin. This requires that

$$u_{nl}(0) = 0. \tag{6.14}$$

For a given l, the eigenenergy is $(2l+1)$-fold degenerate, corresponding to the $2l+1$ possible different valuses of m which varies from $-l$ to l. This degeneracy results from the spherical symmetry of the Hamiltonian. Since $m\hbar$ measures the projection of the orbital angular momentum on the z-axis, the energy levels are not m-dependent.

In order to examine the asymptotic behavior of $u_{nl}(r)$ in the limit $r \to 0$, we assume that in the vicinity of the origin, the interaction potential $V(r)$ has the form

$$V(r) = r^p(a_0 + a_1 r + \cdots), \qquad a_0 \neq 0, \tag{6.15}$$

with $p \geq -1$, which means that the repulsive centrifugal potential, i.e. the first term on the right hand side of Eq. (6.13), is the most singular part of the effective potential.

Since the origin is a regular singular point of the differential equation, we can expand $u_{nl}(r)$ in the vicinity of the origin as

$$u_{nl}(r) = r^s (b_0 + b_1 r + \cdots), \quad b_0 \neq 0. \tag{6.16}$$

Since $u_{nl}(0) = 0$, s should be a positive number. Substituting this expression into the eigen-equation, we find up to the lowest order approximation in r,

$$\frac{\hbar^2}{2\mu} r^{s-2} [l(l+1) - s(s-1)] + a_0 r^{p+s} + \cdots = E r^s + \cdots \tag{6.17}$$

By looking at the coefficient of the lowest power of r, we find that s must satisfy the equation

$$s(s-1) = l(l+1) \tag{6.18}$$

so that $s = l+1$ or $s = -l$. The choice $s = -l$ does not satisfy the condition $u_{nl}(0) = 0$. Thus we should have $s = l+1$ and in the limit $r \to 0$

$$u_{nl}(r) \sim r^{l+1}. \tag{6.19}$$

The corresponding radial function $R_{nl}(r)$ behaves like

$$\boxed{R_{nl}(r \to 0) \sim r^l.} \tag{6.20}$$

This is clearly due to the presence of the centrifugal potential which blocks out the region near the origin when $l \neq 0$.

6.2 HYDROGENIC ATOM

Let us consider a hydrogenic atom containing an atomic nucleus of charge Ze and an electron of charge $-e$ interacting by the Coulomb potential

$$\boxed{V(r) = -\frac{Ze^2}{4\pi\varepsilon_0 r},} \tag{6.21}$$

where r is the distance between the electron and the nucleus. We denote by m_e the mass of the electron and by M the mass of the nucleus. The Hamiltonian of the system reads

$$H = -\frac{\hbar^2}{2M} \nabla_{\mathbf{r}_1}^2 - \frac{\hbar^2}{2m_e} \nabla_{\mathbf{r}_2}^2 + V(|\mathbf{r}_1 - \mathbf{r}_2|). \tag{6.22}$$

6.2.1 HAMILTONIAN IN THE CENTER-OF-MASS FRAMEWORK

Since the interaction potential depends only on the relative coordinate of the two particles, it is convenient to introduce the relative coordinate

$$\mathbf{r} = \mathbf{r}_1 - \mathbf{r}_2, \tag{6.23}$$

and the center-of-mass coordinate

$$\mathbf{R} = \frac{M\mathbf{r}_1 + m_e\mathbf{r}_2}{M + m_e}. \tag{6.24}$$

The Hamiltonian can then be expressed as

$$H = -\frac{\hbar^2}{2(M + m_e)}\nabla_{\mathbf{R}}^2 - \frac{\hbar^2}{2\mu}\nabla_{\mathbf{r}}^2 + V(r), \tag{6.25}$$

where

$$\mu = \frac{Mm_e}{M + m_e} \tag{6.26}$$

is the reduced mass of electron. This separates the relative coordinate from the center-of-mass coordinate.

In the framework of the center-of-mass, we need only solve the Hamiltonian for the relative motion. The corresponding Schrödinger equation reads

$$\boxed{\left(-\frac{\hbar^2}{2\mu}\nabla_{\mathbf{r}}^2 - \frac{Ze^2}{4\pi\varepsilon_0 r}\right)\Psi(\mathbf{r}) = E\Psi(\mathbf{r}).} \tag{6.27}$$

Eq. (6.27) admits a solution of $\Psi(\mathbf{r}) = R_{nl}(r)Y_{lm}(\theta, \phi)$ with $Y_{lm}(\theta, \phi)$ the eigenfunction of $(\mathbf{L}^2, L_z)$. $u_{nl}(r) = rR_{nl}(r)$ is determined by the equation

$$\left[-\frac{\hbar^2}{2\mu}\frac{d^2}{dr^2} + V_l(r)\right]u_{nl}(r) = Eu_{nl}(r), \tag{6.28}$$

where

$$V_l(r) = \frac{l(l+1)\hbar^2}{2\mu r^2} - \frac{Ze^2}{4\pi\varepsilon_0 r}. \tag{6.29}$$

The effective potential $V_{\text{eff}}(r)$ approaches to zero in the large r limit, the eigenstates are unbounded and the corresponding eigenvalues are continuous for $E > 0$.

6.2.2 BOUND STATE SOLUTIONS

In what follows, we focus on the solution of the bound states, for which $E < 0$. It is convenient to introduce the dimensionless quantities

$$z = \frac{\sqrt{-8\mu E}}{\hbar}r, \tag{6.30}$$

$$n = \frac{Ze^2}{4\pi\varepsilon_0\hbar}\sqrt{-\frac{\mu}{2E}}. \tag{6.31}$$

Here n is just a quantum parameter that characterizes the radical function u_{nl}. Below we show that it can only take integer numbers. In terms of these new variables, the eigen-equation becomes

$$\left[\frac{d^2}{dz^2} - \frac{l(l+1)}{z^2} + \frac{n}{z} - \frac{1}{4}\right]u_{nl}(z) = 0. \tag{6.32}$$

In the large z limit, the terms in z^{-1} and z^{-2} become negligible with respect to the constant term, the above equation reduces to

$$\left(\frac{d^2}{dz^2} - \frac{1}{4}\right) u_{nl}(z) = 0. \tag{6.33}$$

The allowed solution is simply given by

$$u_{nl}(z) \sim e^{-z/2}, \quad z \to \infty. \tag{6.34}$$

This, together with the fact that $u_{nl}(z)$ behaves like z^{l+1} in the vicinity of the origin, suggests us to look for a solution of the form

$$u_{nl}(z) = z^{l+1} e^{-z/2} f(z), \tag{6.35}$$

where $f(z)$ is a function of z that satisfies the equation

$$\boxed{\left[z\frac{d^2}{dz^2} + (2l+2-z)\frac{d}{dz} + n - l - 1\right] f(z) = 0.} \tag{6.36}$$

$f(z)$ should be finite in both $z \to 0$ and $z \to \infty$ limits. In this sense, $f(z)$ such defined is regularized.

6.2.3 SOLUTIONS BY SERIES EXPANSION

As $f(z)$ is a regular function of z, we solve Eq. (6.36) by assuming $f(z)$ to have the form

$$f(z) = \sum_{k=0} c_k z^k. \tag{6.37}$$

Substituting Eq. (6.37) into Eq. (6.36), we obtain

$$\sum_{k=0} [(k+2l+2)(k+1)c_{k+1} + (n-l-1-k)c_k] z^k = 0. \tag{6.38}$$

So the coefficients c_k must satisfy the recursion relation

$$c_{k+1} = -\frac{n-l-1-k}{(k+2l+2)(k+1)} c_k. \tag{6.39}$$

If c_k does not vanish at a sufficiently large k, then in the large k limit

$$\frac{c_{k+1}}{c_k} \sim \frac{1}{k} \tag{6.40}$$

which is a ratio for the series of $z^p \exp(z)$ with p a anbitrary finite value since

$$e^z = \sum_k \frac{1}{k!} z^k. \tag{6.41}$$

In this case $u_{nl}(z)$ has an asymptotic form

$$u_{nl}(z) \sim z^{l+1+p} e^{z/2}, \quad z \to \infty, \tag{6.42}$$

which diverges in the limit $z \to \infty$. To avoid this divergency, the series expansion should therefore terminate. This means that $f(z)$ must be a polynomial, not an infinite series summation, in z.

Let k_{max} be the largest power in $f(z)$, we then have

$$n = k_{max} + l + 1. \tag{6.43}$$

$n \geq l + 1$ is a positive integer, called the principal quantum number.

The bound state energy is therefore given by

$$E_n = -\frac{\mu}{2\hbar^2} \left(\frac{Ze^2}{4\pi\varepsilon_0} \right)^2 \frac{1}{n^2} = -\frac{e^2}{4\pi\varepsilon_0 a_\mu} \frac{Z^2}{2n^2}, \tag{6.44}$$

which depends only on the principal quantum number n. a_μ is the Bohr radius corresponding to the reduced mass

$$a_\mu = \frac{m_e}{\mu} a_0, \quad a_0 = \frac{4\pi\varepsilon_0 \hbar^2}{m_e e^2} = 5.29 \times 10^{-11} \text{m}, \tag{6.45}$$

a_0 is the Bohr radius.

The energy level such obtained agrees exactly with that found from the Bohr model. It also agrees the main features of the experimental spectrum. The agreement, however, is not perfect and various corrections, such as the fine structure arising form relativistic effects, the electron spin, the Lamb shift, and the hyperfine structure due to nuclear effects, should be taken into account in order to explain the detailed experimental spectrum.

Since n may take on all integral values from 1 to infinity, there are infinite number of discrete bound state levels. This is due to the fact that the magnitude of the Coulomb potential falls off slowly at large r.

Since the energy level E_n does not depend on the magnetic quantum numbers l and m, the eigenenergy is therefore degenerate. For each given n, l may take on the values from 0 to $n - 1$, and for each value l, the magnetic quantum number m may take any one of the $2l + 1$ values from $-l$ to l. For each allowed (n, l, m), the electron can be in any one of the two spin states. Thus the total degeneracy of the bound-state energy level E_n is

$$2 \sum_{l=0}^{n-1} (2l + 1) = 2n^2. \tag{6.46}$$

In a given eigenstate of the Schrödinger equation characterized by the quantum numbers (n, l, m), the expectation of the kinetic energy is related to the expectation of the potential energy by the virial theorem presented in Section 3.4.1,

$$\langle T \rangle_{nlm} = -\frac{1}{2} \langle V(r) \rangle_{nlm}, \tag{6.47}$$

where

$$T = -\frac{\hbar^2}{2\mu}\nabla_r^2. \tag{6.48}$$

Thus the total energy in this state is

$$E_n = \langle T\rangle_{nlm} + \langle V\rangle_{nlm} = \frac{1}{2}\langle V\rangle_{nlm} = -\langle T\rangle_{nlm}. \tag{6.49}$$

It indicates that the expectation value of the kinetic energy as well as that of the potential energy does not depend on the magnetic quantum numbers l and m. From the above equation, we find the expectation value of $1/r$ to be

$$\left\langle\frac{1}{r}\right\rangle_{nlm} = -\frac{8\pi\varepsilon_0}{Ze^2}E_n = \frac{Z}{n^2a_\mu}, \tag{6.50}$$

which is also independent of l and m.

6.2.4 RADIAL WAVE FUNCTION

The radial function is now given by

$$R_{nl}(r) = \frac{1}{r}u_{nl}(z) = \frac{1}{r}z^{l+1}e^{-z/2}f(z), \tag{6.51}$$

where z is n dependent

$$z = \frac{r\sqrt{-8\mu E}}{\hbar} = \frac{2rZ}{na_\mu} \tag{6.52}$$

and $f(z)$ is a polynomial of degree $k_{max} = n - l - 1$ defined by Eq. (6.37) whose coefficients are determined by the recursion formula

$$c_{k+1} = -\frac{n-l-1-k}{(k+2l+2)(k+1)}c_k. \tag{6.53}$$

The volume element in spherical coordinate is

$$d\mathbf{r} = r^2\sin\theta dr d\theta d\phi, \tag{6.54}$$

so the normalization condition of the wave function becomes

$$\int d\mathbf{r}|\Phi(\mathbf{r})|^2 = \int |R(r)|^2 r^2 dr = 1. \tag{6.55}$$

The ground state is the case $n = 1$, $l = m = 0$, the radial wave function, after normalization is

$$R_{10}(r) = \frac{2Zc_0}{a_\mu}e^{-rZ/a_\mu}. \tag{6.56}$$

Normalizing it, we find that

$$1 = \int_0^\infty drr^2R_{10}^2(r) = \frac{4Z^2c_0^2}{a_\mu^2}\int_0^\infty drr^2e^{-2rZ/a_\mu} = \frac{a_\mu c_0^2}{Z}, \tag{6.57}$$

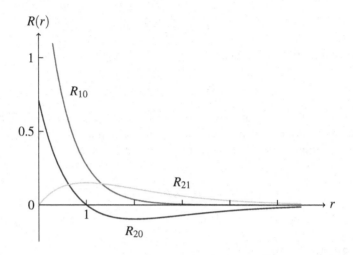

Figure 6.1   Radial wave function $R(r)$ [in unit of $(Z/a_\mu)^{3/2}$] versus r [in unit of $2a_\mu/Z$] for R_{10}, R_{20} and R_{21}.

so that $c_0 = \sqrt{Z/a_\mu}$. Thus we have

$$R_{10}(r) = 2\frac{Z^{3/2}}{a_\mu^{3/2}}e^{-rZ/a_\mu}. \tag{6.58}$$

If $n = 2$ and $l = 0$, $k_{max} = 1$ and

$$c_1 = -\frac{1}{2}c_0. \tag{6.59}$$

The radial wave function is given by

$$R_{20}(r) = \frac{Z^{3/2}}{\sqrt{2}a_\mu^{3/2}}\left(1 - \frac{rZ}{2a_\mu}\right)e^{-rZ/2a_\mu}. \tag{6.60}$$

R_{20} has a node at $r = 2a_\mu/Z$.
If $n = 2$ and $l = 1$,

$$R_{21}(r) = \frac{\sqrt{6}Z^{5/2}}{12a_\mu^{5/2}}re^{-rZ/2a_\mu}. \tag{6.61}$$

The r-dependence of (R_{10}, R_{20}, R_{21}) is shown in Fig. 6.1. In general, $f(z)$ may be expressed in terms of associated Laguerre polynomials L_q^p as

$$\boxed{R_{nl}(r) = Nz^l e^{-z/2}L_{n+l}^{2l+1}(z),} \tag{6.62}$$

where N is the normalization constant and L_q^p is related to the Laguerre polynomials L_q by the formula

$$L_q^p(z) = \frac{d^p}{dz^p} L_q(z), \tag{6.63}$$

and

$$L_q(z) = e^z \frac{d^q}{dz^q} \left(z^q e^{-z} \right). \tag{6.64}$$

6.2.5 RYDBERG FORMULA

From the solution of the Schrödinger equation for the hydrogenic atom, we know that the eigenenergy of electron is determined purely by the radial quantum number n. An electron can in principle make a transition from one eigenenergy state to another by either absorbing or emitting a photon with a proper energy. The energy of absorbed or emitted photon should equal the energy change from the two eigenenergies of the electron, namely

$$\hbar\omega = \frac{\mu e^4}{2(4\pi\varepsilon_0\hbar)^2} \left| \frac{1}{n_i^2} - \frac{1}{n_f^2} \right|, \tag{6.65}$$

where n_i and n_f are the initial and final radial quantum numbers of the electron, respectively. If $n_f < n_i$, it is to emit a photon. Otherwise, it is to absorb a photon.

From the above formula, we find the inverse of the wavelength of the photon, which is the quantity measured in optical spectra, to be

$$\lambda^{-1} = \frac{\omega}{2\pi c} = R \left| \frac{1}{n_i^2} - \frac{1}{n_f^2} \right|, \tag{6.66}$$

where R is the Rydberg constant

$$R = \frac{\mu e^4}{(4\pi)^3 c\varepsilon_0^2\hbar^3} \tag{6.67}$$

Eq. (6.66) is known as the Rydberg formula. This formula was discovered empirically in the nineteenth century. It was first interpreted by Niels Bohr using his model of the atom in 1913. In the emission spectrum, depending on the final radial quantum number $n_f < n_i$, the spectral lines can fall into different optical bands. The greater the difference in the principal quantum numbers, the higher the energy of the electromagnetic emission. For example,

1. The spectral lines resulting from the transitions to the ground state, $n_f = 1$, are in the ultraviolet band. This set of spectral lines is called the Lyman series.
2. The spectral lines resulting from the transitions to the first excited level, $n_f = 2$, are in the visible band, This set of spectral lines is called the Balmer series.

3. The spectral lines resulting from the transitions to the second excited level, $n_f = 3$, are in the infrared band. This set of spectral lines is called the Paschen series.

6.3 PARTIAL WAVE METHOD

Here we consider the scattering of a beam of particles by a spherically symmetric potential, as illustrated by Fig. 6.2.

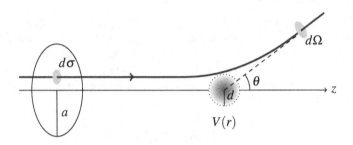

Figure 6.2 Scattering of particles incident in the area $d\sigma$ into the solid angle $d\Omega$ by a short-ranged central potential $V(r)$.

To solve this scattering problem analytically, the following assumptions or approximations are adopted:

1. The beam is switched on for times long compared to the time of interaction of the particles with the scattering potential so that only a steady state, whose wave function is governed by the stationary Schrödinger equation (6.1), needs to be solved.
2. The potential is short ranged with a finite force range d. Thus in the large distance limit, $r \to \infty$, the particles behave like free particles. Hence $V(r) = 0$ if $r > d$.
3. The radius a of the window through which the incident particles are emitting into the scattering area is significantly larger than the potential range d, namely $a \gg d$, so that the incident wave could be initially simulated by a plane wave propagating along the z-axis,

$$\Psi_{\text{inc}}(\mathbf{r}) = e^{ikz} = e^{ikr\cos\theta}, \qquad (r \to \infty) \qquad (6.68)$$

where θ is the polar angle and k is the wave vector of the incident particle

$$k = \frac{\sqrt{2\mu E}}{\hbar}. \qquad (6.69)$$

After encountering the scattering potential, the particles produce an outgoing spherical wave in the large distance limit. As the system is azimuthal symmetric, the trajectory of each particle should remain in a place with a fixed azimuthal angle in the classical picture. The scattering wave function, Ψ_{sc}, is expected to be described by a spherically outgoing wave packet whose probability current density is pointing asymptotically along the radial direction. From the probability conservation, the scattering wave is anticipated to decay linearly with the inverse of the distance r in the $r \rightarrow \infty$ limit

$$\Psi_{sc}(\mathbf{r}) = f(\theta, k)\frac{e^{ikr}}{r}, \qquad (r \rightarrow \infty), \tag{6.70}$$

independent on the azimuthal angle φ. $f(\theta, k)$ is the scattering amplitude.

In the large r region, the wave function $\Psi(\mathbf{r})$ should describe both the incident state of particles and the state of particles being scattered, so it should be a linear superposition of Ψ_{inc} and Ψ_{sc}

$$\boxed{\Psi(\mathbf{r}) = \Psi_{inc}(\mathbf{r}) + \Psi_{sc}(\mathbf{r}), \qquad (r \rightarrow \infty).} \tag{6.71}$$

It is simple to verify that Ψ_{sc} satisfies the Schrödinger equation for the free particle in the limit $r \rightarrow \infty$

$$-\frac{\hbar^2}{2\mu}\nabla^2\Psi_{sc} = E\Psi_{sc}. \tag{6.72}$$

Using the formula

$$\nabla = \hat{r}\frac{\partial}{\partial r} + \hat{\theta}\frac{1}{r}\frac{\partial}{\partial\theta} + \hat{\varphi}\frac{1}{r\sin\theta}\frac{\partial}{\partial\varphi} \tag{6.73}$$

it can be further shown that the probability current of the scattering function Ψ_{sc} is a radial vector in the large r limit

$$\boxed{\mathbf{j}_{sc} = \frac{i\hbar}{2\mu}\left(\Psi_{sc}\nabla\Psi_{sc}^* - \Psi_{sc}^*\nabla\Psi_{sc}\right) \xrightarrow[r\rightarrow\infty]{} \frac{\hbar k}{\mu}\frac{|f(k,\theta)|^2}{r^2}\hat{r}.} \tag{6.74}$$

6.3.1 PARTIAL WAVE EXPANSION

Partial wave analysis is a useful method for determining the scattering amplitude $f(\theta, k)$. It solves the scattering problem by decomposing the wave function into its constituent angular momentum components and determining the coefficient of each component using the boundary condition at large distances.

This problem is soluble because only the asymptotic expression of the wave function at large distances is needed, which allows the scattering amplitude to be represented using the scattering phase shifts, without knowing the detailed expression of $V(r)$. However, to find the values of these phase shifts, one has to solve the Schrödinger equation in the whole space and match the solution with the asymptotic wave function of the incident and scattering waves in the $r \rightarrow \infty$ limit.

In case of a spherically symmetric potential, the wave function can be expanded in spherical harmonics which reduce to Legendre polynomials because of azimuthal symmetry

$$\Psi(\mathbf{r}) = \sum_{l=0}^{\infty} R_l(r, k)P_l(\cos\theta), \tag{6.75}$$

where the radial function is governed by the equation

$$\left[\frac{d^2}{dr^2} + \frac{2}{r}\frac{d}{dr} - \frac{l(l+1)}{r^2} - U(r) + k^2 \right] R_l(r,k) = 0, \qquad (6.76)$$

with

$$U(r) = \frac{2\mu}{\hbar} V(r). \qquad (6.77)$$

In the region where the scattering potential $V(r)$ becomes zero (or negligible), the equation becomes

$$\left[\frac{d^2}{dr^2} + \frac{2}{r}\frac{d}{dr} - \frac{l(l+1)}{r^2} + k^2 \right] R_l(r,k) = 0. \qquad (6.78)$$

By setting $x = rk$, it further reduces to

$$\left[\frac{d^2}{dx^2} + \frac{2}{x}\frac{d}{dx} - \frac{l(l+1)}{x^2} + 1 \right] R_l(x) = 0. \qquad (6.79)$$

Hence $R_l(r,k) = R_l(kr)$ is just a function of kr in the large r limit.

For a given l, the general solution of Eq. (6.79) is a linear superposition of the spherical Bessel and Neumann functions $j_l(x)$ and $n_l(x)$

$$R_l(x) = b_l j_l(x) + c_l n_l(x), \qquad (6.80)$$

with b_l and c_l the normalization constants. In the limit $r \to \infty$, $j_l(x)$ and $n_l(x)$ are asymptotically given by

$$j_l(x) \xrightarrow[x \to \infty]{} \frac{1}{x}\sin\left(x - \frac{l\pi}{2}\right), \qquad (6.81)$$

$$n_l(x) \xrightarrow[x \to \infty]{} -\frac{1}{x}\cos\left(x - \frac{l\pi}{2}\right). \qquad (6.82)$$

Hence, $R_l(x)$ asymptotically behaves as

$$R_l(x) \xrightarrow[x \to \infty]{} \frac{1}{x}\left[b_l \sin\left(x - \frac{l\pi}{2}\right) - c_l \cos\left(x - \frac{l\pi}{2}\right) \right]$$
$$= \frac{a_l}{x}\sin\left(x - \frac{l\pi}{2} + \delta_l\right), \qquad (6.83)$$

where

$$a_l = \sqrt{b_l^2 + c_l^2}, \qquad (6.84)$$

and

$$\delta_l = -\tan^{-1}\frac{c_l}{b_l} \qquad (6.85)$$

is the phase shift of the l'th partial wave caused by the scattering potential, which should be determined by solving Eq. (6.76).

Therefore, the radial function at the large distances could be represented using the phase shifts as

$$R_l(\theta,k) = b_l [j_l(kr) - \tan \delta_l n_l(kr)], \qquad (r > d). \tag{6.86}$$

6.3.2 SCATTERING AMPLITUDE

We now determine the scattering amplitude $f(\theta,k)$ by matching the wave function obtained by solving Eq. (6.79) with the asymptotical incident and scattering wave function given by Eq. (6.71) in the large r limit. For doing this, we first expand Ψ_{inc} using the Legendre polynomials

$$e^{ikr\cos\theta} = \sum_{l=0} (2l+1)i^l j_l(kr)P_l(\cos\theta). \tag{6.87}$$

In the large r limit, it becomes

$$e^{ikr\cos\theta} = \sum_{l=0} \frac{(2l+1)i^l}{kr} \sin\left(kr - \frac{l\pi}{2}\right) P_l(\cos\theta). \tag{6.88}$$

Similarly, the scattering amplitude $f(\theta,k)$ can be expanded as

$$f(\theta,k) = \sum_{l=0} f_l(k)P_l(\cos\theta). \tag{6.89}$$

Thus the asymptotic form of the radial function R_l is

$$R_l(r,k) \xrightarrow[r\to\infty]{} \frac{(2l+1)i^l}{kr} \sin\left(kr - \frac{l\pi}{2}\right) + \frac{1}{r} e^{ikr} f_l(k). \tag{6.90}$$

Equating it with Eq. (6.83), we obtain

$$\frac{a_l}{k} \sin\left(kr - \frac{l\pi}{2} + \delta_l\right) = \frac{(2l+1)i^l}{k} \sin\left(kr - \frac{l\pi}{2}\right) + e^{ikr} f_l(k). \tag{6.91}$$

This leads to

$$a_l = i^l(2l+1)e^{i\delta_l}, \tag{6.92}$$

$$f_l = \frac{2l+1}{2ik}\left(e^{2i\delta_l} - 1\right) = \frac{2l+1}{k} e^{i\delta_l} \sin \delta_l. \tag{6.93}$$

So the scattering amplitude is related to the phase shifts by the formula

$$f(\theta,k) = \sum_{l=0} (2l+1)\frac{e^{i\delta_l}\sin\delta_l}{k} P_l(\cos\theta). \tag{6.94}$$

6.3.3 SCATTERING CROSS SECTION

The differential scattering cross section $d\sigma/d\Omega$ is defined as the ratio of the number of particles scattered into the direction (θ,φ), per unit time, per unit solid angle, divided by the incident flux per unit area. From Eq. (6.74), we know that the number of particles crossing a solid angle $d\Omega$ per unit time is

$$dN = (\mathbf{j}_{sc}\cdot\hat{r})\,r^2 d\Omega = \frac{\hbar k}{\mu}|f(k,\theta)|^2\,d\Omega. \tag{6.95}$$

It must equal the number of incident particles through an area $d\sigma$ per unit time, i.e.

$$dN = \frac{\hbar k}{\mu}d\sigma. \tag{6.96}$$

Hence the differential cross section equals the absolute square of the scattering amplitude

$$\boxed{\frac{d\sigma}{d\Omega} = |f(\theta,k)|^2.} \tag{6.97}$$

The total cross section is the integral of $d\sigma/d\Omega$ over all the solid angles

$$\sigma = \int |f(\theta,k)|^2 d\Omega = 2\pi \int_{-1}^{1} d(\cos\theta)|f(\theta,k)|^2. \tag{6.98}$$

Using the orthogonality condition of the Legendre polynmials

$$\int_{-1}^{1} dx P_l(x)P_{l'}(x) = \frac{2}{2l+1}\delta_{l,l'}, \tag{6.99}$$

we find that

$$\boxed{\sigma = \sum_{l=0}^{\infty} \frac{4\pi(2l+1)}{k^2}\sin^2\delta_l.} \tag{6.100}$$

6.3.4 HARD-SPHERE SCATTERING

The above discussion indicates that the phase shift is the key parameter in characterizing the scattering strength at each partial wave channel. Now we take the hard-sphere potential as an example to demonstrate how the phase shifts are determined.

The potential reads

$$V(r) = \begin{cases} 0 & r>d, \\ +\infty & r<d. \end{cases} \tag{6.101}$$

Since the particle cannot penetrate into the interior of this hard sphere potential, the wave function at the boundary $r=d$ must vanish

$$\Psi(r=d) = 0. \tag{6.102}$$

It further requires the radial function of each partial wave to be zero at $r = d$, i.e.

$$R_l(d,k) = b_l \left[j_l(kd) - \tan \delta_l n_l(kd) \right] = 0, \tag{6.103}$$

from which it follows that

$$\tan \delta_l = \frac{j_l(kd)}{n_l(kd)}. \tag{6.104}$$

In the long wavelength limit, $kd \ll 1$, the Bessel and Neumann functions behave asymptotically as

$$j_l(kd) \xrightarrow[kd \to 0]{} \frac{(kd)^l}{(2l+1)!!}, \tag{6.105}$$

$$n_l(kd) \xrightarrow[kd \to 0]{} -\frac{(2l-1)!!}{(kd)^{l+1}}, \tag{6.106}$$

the corresponding phase shift is

$$\tan \delta_l \approx -\frac{(kd)^{2l+1}}{(2l+1)!!(2l-1)!!}, \tag{6.107}$$

where

$$(2l+1)!! = 1 \cdot 3 \cdot 5 \cdots (2l+1) \quad \text{and} \quad (-1)!! = 1. \tag{6.108}$$

So $\tan \delta_l$ decays quickly with l, and the phase shift is dominated by the s-wave ($l = 0$) component and

$$\delta_l \approx -kd \delta_{l,0}. \tag{6.109}$$

Hence the total cross section is

$$\sigma \xrightarrow[kd \ll 1]{} 4\pi d^2, \tag{6.110}$$

independent on k. The scattering cross section is the total surface area of the sphere, which is four times of the geometric cross section of the sphere πd^2.

6.4   SUPERSYMMETRIC QUANTUM MECHANICS APPROACH*

Supersymmetric quantum mechanics is an elegant approach for solving a series of one-dimensional eigen-problems with pairwise potentials which are related to each other by the so-called superpotentials. Each Hamiltonian has a partner Hamiltonian which has the same energy spectrum, except for the zero energy eigenstates. This fact can be exploited to deduce the eigenstate spectrum from one system to another. It is analogous to the original description of supersymmetry between bosons and fermions. We can imagine a "bosonic Hamiltonian", whose eigenstates are the various bosons. The supersymmetric partner of this Hamiltonian would be "fermionic". Each boson would have a fermionic partner of equal energy.

In one dimension, there is an intrinsic connection between the bound state wave function and the potential. In fact, once a bound state wave function is exactly known, the potential is completely fixed. For this propose, the ground state wave function is more often used because it has no nodes.

Let us start from an arbitrary Hamiltonian

$$H = -\frac{\hbar^2}{2\mu}\frac{\partial^2}{\partial x^2} + V(x). \tag{6.111}$$

Let $\Psi_0(x)$ and E_0 be the ground state wave function and the corresponding eigen-energy. If E_0 is non zero, we can always redefine the Hamiltonian by a constant energy shift for the potential

$$H_1 = -\frac{\hbar^2}{2\mu}\frac{\partial^2}{\partial x^2} + V_1(x), \tag{6.112}$$

$$V_1(x) \equiv V(x) - E_0, \tag{6.113}$$

so that the ground state energy equals zero. It follows that the Schröinger equation for the ground state takes the form

$$\left[-\frac{\hbar^2}{2\mu}\frac{\partial^2}{\partial x^2} + V_1(x)\right]\Psi_0(x) = 0, \tag{6.114}$$

and the potential $V_1(x)$ is purely determined by the ground state wave function

$$V_1(x) = \frac{\hbar^2}{2\mu}\frac{\Psi_0''(x)}{\Psi_0(x)}. \tag{6.115}$$

This expression actually offers a simple scheme to factorize the Hamiltonian as a product of an operator Q and its hermitian conjugate $Q^\dagger$:

$$\boxed{H_1 = Q^\dagger Q.} \tag{6.116}$$

Q and $Q^\dagger$ are known as supercharges. They are defined in terms of a superpotential $W(x)$ as

$$\boxed{\begin{aligned} Q &= \frac{\hbar}{\sqrt{2\mu}}\frac{\partial}{\partial x} + W(x), \\[2mm] Q^\dagger &= -\frac{\hbar}{\sqrt{2\mu}}\frac{\partial}{\partial x} + W(x). \end{aligned}} \tag{6.117}$$

These operators can be regarded as a generalization of the annihilation and creation operators for the quantum harmonic oscillator. The potential is related to the super-potential by the formula

$$\boxed{V_1(x) = -\frac{\hbar}{\sqrt{2\mu}}\frac{\partial W}{\partial x} + W^2.} \tag{6.118}$$

The superpotential is determined by solving the ground state eigen-equation

$$Q^\dagger Q \Psi_0(x) = 0. \tag{6.119}$$

This is equivalent to solving the equation

$$Q \Psi_0(x) = \frac{\hbar}{\sqrt{2\mu}} \frac{\partial}{\partial x} \Psi_0(x) + W(x) \Psi_0(x) = 0. \tag{6.120}$$

The solution to this equation is

$$\boxed{W(x) = -\frac{\hbar}{\sqrt{2\mu}} \frac{d}{dx} \ln \Psi_0(x).} \tag{6.121}$$

Now we define a partner Hamiltonian H_2 from H_1 by reversing the order of Q and $Q^\dagger$

$$\boxed{H_2 = QQ^\dagger = -\frac{\hbar^2}{2\mu} \frac{\partial^2}{\partial x^2} + V_2(x),} \tag{6.122}$$

this yields a new potential

$$\boxed{V_2(x) = \frac{\hbar}{\sqrt{2\mu}} \frac{\partial W}{\partial x} + W^2.} \tag{6.123}$$

The potentials $V_1(x)$ and $V_2(x)$ are known as supersymmetric partner potentials.

Both H_1 and H_2 are positive semidefinite. However, their eigenstates, except the ground state of H_1, are inherently related to each other:

1. If $\left| \Psi_n^{(1)} \right\rangle$ is an eigenstate of H_1 with the eigenvalue $E_n > 0$

$$H_1 \left| \Psi_n^{(1)} \right\rangle = Q^\dagger Q \left| \Psi_n^{(1)} \right\rangle = E_n \left| \Psi_n^{(1)} \right\rangle, \tag{6.124}$$

then $\left| \Psi_n^{(2)} \right\rangle \propto Q \left| \Psi_n^{(1)} \right\rangle$ is an eigenstate of H_2 with the same eigenvalue

$$H_2 \left(Q \left| \Psi_n^{(1)} \right\rangle \right) = QQ^\dagger Q \left| \Psi_n^{(1)} \right\rangle = E_n \left(Q \left| \Psi_n^{(1)} \right\rangle \right). \tag{6.125}$$

The normalizated eigenfunction $\left| \Psi_n^{(2)} \right\rangle$ is

$$\boxed{\left| \Psi_n^{(2)} \right\rangle = \frac{1}{\sqrt{\left\langle \Psi_n^{(1)} \left| Q^\dagger Q \right| \Psi_n^{(1)} \right\rangle}} Q \left| \Psi_n^{(1)} \right\rangle = \frac{1}{\sqrt{E_n}} Q \left| \Psi_n^{(1)} \right\rangle.} \tag{6.126}$$

2. Similarly, if $\left|\Psi_n^{(2)}\right\rangle$ is an eigenstate of H_2 with the eigenvalue $E_n > 0$

$$H_2\left|\Psi_n^{(2)}\right\rangle = QQ^\dagger\left|\Psi_n^{(2)}\right\rangle = E_n\left|\Psi_n^{(2)}\right\rangle, \qquad (6.127)$$

then $\left|\Psi_n^{(1)}\right\rangle \propto Q^\dagger\left|\Psi_n^{(2)}\right\rangle$ is an eigenstate of H_1 with the same eigenvalue

$$H_2\left(Q^\dagger\left|\Psi_n^{(2)}\right\rangle\right) = Q^\dagger QQ^\dagger\left|\Psi_n^{(2)}\right\rangle = E_n\left(Q^\dagger\left|\Psi_n^{(2)}\right\rangle\right). \qquad (6.128)$$

The normalizated eigenfunction is

$$\left|\Psi_n^{(1)}\right\rangle = \frac{1}{\sqrt{\left\langle\Psi_n^{(2)}\left|QQ^\dagger\right|\Psi_n^{(2)}\right\rangle}}Q^\dagger\left|\Psi_n^{(2)}\right\rangle = \frac{1}{\sqrt{E_n}}Q^\dagger\left|\Psi_n^{(2)}\right\rangle. \qquad (6.129)$$

Thus all the eigenstates of H_2 have their supersymmetric partners in H_1 with the same eigenvalues. The ground state of H_1 has no supersymmetric partner because it is annihilated by Q. From the eigen-solutions of H_1 we can determine the eigenfunctions of H_2 using Q. Similarly, from the eigen-solutions of H_2, we can determine all the eigenfunctions of H_1 except for the ground state. This supersymmetric scheme offers a convenient approach to use the eigen-solution in one potential to solve the eigen-problem in its partner potential.

The above discussion indicates that two Hamiltonians, H_1 and H_2, form a pair of supersymmetric partners when the difference between their potentials is related to the ground state wave function $\Psi_0^{(1)}(x) = \Psi_0(x)$ of H_1 by the equation

$$V_2(x) - V_1(x) = \frac{\sqrt{2}\hbar}{\sqrt{\mu}}\frac{\partial W}{\partial x} = -\frac{\hbar^2}{\mu}\frac{d^2}{dx^2}\ln\Psi_0(x). \qquad (6.130)$$

Similarly, one can use the ground state of H_2, $\Psi_1^{(2)}(x)$, to generate a new potential $V_3(x)$, which is a supersymmetric partner of $V_2(x)$

$$V_3(x) - V_2(x) = -\frac{\hbar^2}{\mu}\frac{d^2}{dx^2}\ln\Psi_1^{(2)}(x). \qquad (6.131)$$

The corresponding Hamiltonian is

$$H_3 = -\frac{\hbar^2}{2\mu}\frac{\partial^2}{\partial x^2} + V_3(x). \qquad (6.132)$$

Again, each eigenstate of H_2, except the ground state $\Psi_1^{(2)}$, has its supersymmetric partner with the same eigenenergy.

Repeat the above procedure iteratively, we generate a Hamiltonian hierarchy with the property that the n'th member of the hierarchy has the same eigenvalue spectrum as the first member H_1, except for the lowest $(n-1)$ eigenvalues of H_1. The n'th excited state of H_1 is degenerate with the ground state of H_{n+1}, and the corresponding wave functions are related to each by the supercharge operators.

6.4.1 SUPERSYMMETRIC SOLUTION OF THE HYDROGENIC ATOM

As an appealing example, we now use the approach of supersymmetric quantum mechanics to solve the bound state eigen-spectrum of the hydrogen atom.[1] This approach avoids the algebra details of the usual polynomial expansion and demonstrates in an elegant manner the solution of hydrogenic atoms.

As revealed by Eq. (6.28), the radial wave function of a hydrogenic atom $u_{nl}(r) = rR_{nl}(r)$ is determined by the Schrödinger equation in an effective one-dimensional potential

$$H_l u_{nl}(r) = E_{nl} u_{nl}(r), \qquad (r > 0) \tag{6.133}$$

where H_l is the Hamiltonian for a given angular momentum l

$$H_l = -\frac{\hbar^2}{2\mu} \frac{d^2}{dr^2} + V_l(r) \tag{6.134}$$

and $V_l(r)$ is the effective potential defined by Eq. (6.29), i.e.

$$V_l(r) = \frac{l(l+1)\hbar^2}{2\mu r^2} - \frac{Ze^2}{4\pi\varepsilon_0 r}. \tag{6.135}$$

Using the properties of $u_{nl}(r)$ in the following two limits

$$u_{nl}(r \to 0) \sim r^{l+1}, \tag{6.136}$$

$$u_{nl}(r \to \infty) \sim \exp\left(-\frac{\sqrt{2\mu |E_{nl}|}}{\hbar} r\right), \tag{6.137}$$

and the fact that the ground state wave function of H_l must be nodeless, it is simple to show that

$$u_{n_0 l}(r) = C_{n_0 l} r^{l+1} \exp\left(-\frac{\sqrt{2\mu |E_{n_0 l}|}}{\hbar} r\right), \tag{6.138}$$

is the ground state of H_l with the eigenenergy

$$E_{n_0 l} = -\frac{\mu}{2\hbar^2} \left(\frac{Ze^2}{4\pi\varepsilon_0}\right)^2 \frac{1}{n_0^2}, \tag{6.139}$$

and

$$n_0 = l + 1 \tag{6.140}$$

is an integer quantum number labelling this state. So the eigenstates of $(H_0, H_1, \cdots H_l, \cdots)$ start from $n = (1, 2, \cdots, l+1, \cdots)$, respectively.

---

[1] A. Valance, T. J. Morgan, H. Bergeron, Eigensolution of the Coulomb Hamiltonian via supersymmetry, American Journal of Physics, **58**, 487 (1990).

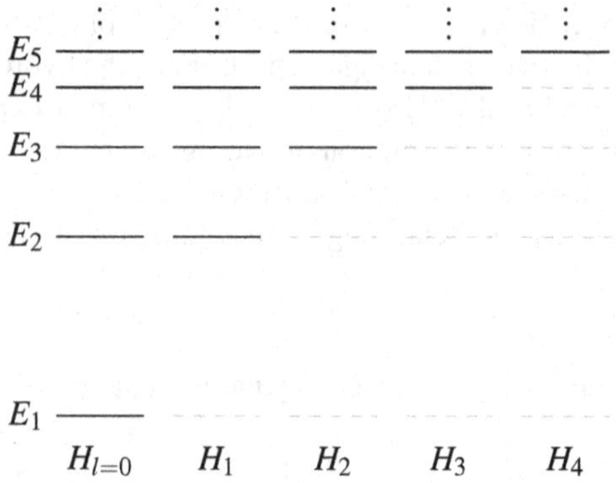

Figure 6.3 Schematic illustration of the supersymmetric hierarchy of the eigenenergies of the hydrogenic atom.

It is straightforward to show that $V_{l+1}(r)$ is the supersymmetric partner potential of $V_l(r)$:

$$\boxed{V_{l+1}(r) - V_l(r) = -\frac{\hbar^2}{\mu}\frac{d^2}{dr^2}\ln u_{n_0 l}(r) = \frac{(l+1)\hbar^2}{\mu r^2}.}$$
(6.141)

So all the bound state eigenvalues of H_{l+1} should be the same as the corresponding eigenvalues of H_l

$$E_{n,l+1} = E_{n,l}, \qquad n > n_0 = l+1.$$
(6.142)

The above derivation implies that all the radial Hamiltonians H_l ($l = 0, 1, 2 \cdots$) actually form a supersymmetric Hamiltonian hierarchy. H_l should have the same energy spectrum as $H_{l=0}$, except for the missing l-lowest eigenvalues of H_0. In particular, the n'th eigenstate of H_0 is degenerate with the ground state of $H_{l=n-1}$ and

$$\boxed{E_{n,0} = E_{n,1} = \cdots = E_{n,n-1} = E_n,}$$
(6.143)

where

$$E_n = -\frac{\mu}{2\hbar^2}\left(\frac{Ze^2}{4\pi\varepsilon_0}\right)^2\frac{1}{n^2}$$
(6.144)

is just the eigenenergy of the hydrogenic atom obtained in Section 6.2 by directly solving the eigen-equation using the polynomial expansion. Fig. 6.3 shows schematically the supersymmetric structure of the energy spectrum of the hydrogenic atom.

Furthermore, the eigenfunctions of H_l and H_{l+1} are related to each other by the superpotential $W_l(r)$, which in turn is determined by the ground state wave function

$u_{l+1,l}(r)$ of H_l

$$
\begin{aligned}
W_l(r) &= -\frac{\hbar}{\sqrt{2\mu}} \frac{d}{dr} \ln u_{l+1,l}(r) \\
&= -\frac{\hbar}{\sqrt{2\mu}} \left(\frac{l+1}{r} - \frac{\sqrt{2\mu \left| E_{l+1,l} \right|}}{\hbar} \right) \\
&= -\frac{\hbar}{\sqrt{2\mu}} \left[\frac{l+1}{r} - \frac{Z}{a_\mu (l+1)} \right].
\end{aligned}
\tag{6.145}
$$

The corresponding supercharge operators are

$$
Q_l = -\frac{\hbar}{\sqrt{2\mu}} \frac{\partial}{\partial r} + W_l(r),
\tag{6.146}
$$

$$
Q_l^\dagger = \frac{\hbar}{\sqrt{2\mu}} \frac{\partial}{\partial r} + W_l(r).
\tag{6.147}
$$

Starting from the ground state wave function of H_l, we can find all the eigenfunctions of the Hamiltonians with $n = l+1$, from H_{l-1} to H_0 by recursively applying the supercharge operators to the eigenfunction obtained in the previous step

$$
u_{n,l-1}(r) = \frac{1}{\sqrt{E_n}} Q_{l-1}^\dagger u_{n,l}(r),
\tag{6.148}
$$

$$
u_{n,l-2}(r) = \frac{1}{\sqrt{E_n}} Q_{l-2}^\dagger u_{n,l-1}(r),
\tag{6.149}
$$

$$
\cdots = \cdots
$$

$$
u_{n,0}(r) = \frac{1}{\sqrt{E_n}} Q_0^\dagger u_{n,1}(r).
\tag{6.150}
$$

This allows us to find all the eigenfunctions of the radial Hamiltonian of the hydrogenic atom just by utilizing the supersymmetric properties of the Hamiltonian.

6.5 PROBLEMS

1. Show that a particle in a spherical potential well

$$V(\mathbf{r}) = \begin{cases} -V_0, & r < a, \\ 0, & r > a, \end{cases}$$

has a bound state only when V_0 is above a threshold.

2. Sudden approximation: an atom of tritium ^{3}H is in its ground state. If it suddenly decays into a ^{3}He$^+$ ion by emitting a fast electron from the nucleus without perturbing the extranuclear electron.

   a. Find the probability that the resulting He ion will be left in a $n = 1$ and $l = 0$ state.

   b. Find the probability that it will be left in a $n = 2$ and $l = 0$ state.

   c. Find the probability that the ion will be left in all $l > 0$ state?

3. Optical Selection Rule: the probability for a transition between two states with quantum numbers (n_i, l_i, m_i) and (n_f, l_f, m_f) in a Hydrogen atom induced by an electric dipole interaction

$$H_{\text{dipole}} = -e\mathbf{r} \cdot \mathbf{E}$$

   is proportional to the modulus square of the matrix element of H_{dipole} between these two states

$$\langle n_f, l_f, m_f | H_{\text{dipole}} | n_i, l_i, m_i \rangle,$$

   where $\mathbf{r}$ is the vector position operator of the electron and $\mathbf{E}$ is an applied electric field.

   Show that the transition probability is zero for the $2s \to 1s$ transition, but nonzero for the $2p \to 1s$ transition.

4. A particle is trapped inside a ball of radius R_0. If the particle is in the ground state, how large is the force the particle acts on the wall of the ball?

5. An electron is in an eigenstate $|nlm\rangle$ of a hydrogenic atom with the potential

$$V(r) = -\frac{Ze^2}{r}$$

   Calculate the expectation values of $1/r$ and $1/r^2$ in this eigenstate.

6. Using the Heisenberg uncertainty principle, estimate the ground state energy of electron in a Hydrogen atom.

7. Calculate the value of $\Delta x \Delta p_x$ for the electron in the ground state of Hydrogen atom.

8. Optical theorem: For a short-ranged spherical scattering potential, show that the total cross section

$$\sigma = \frac{4\pi}{k} \text{Im} f(\theta = 0, k),$$

   $f(\theta, k)$ is the scattering amplitude. This relation is referred to as the optical theorem.

9. Consider the scattering of a particle by a spherical potential well

$$V(r) = \begin{cases} -V_0 & r < d \\ 0 & r > d \end{cases} .$$

where $\theta(r)$ is the step function. Calculate the phase shift δ_l. Assuming $kd \ll 1$ so that only the s-wave ($l = 0$) phase shift contributes significantly, calculate the total cross section. k is the wave vector of the incident particle.

7 Identical Particles

7.1 PERMUTATION SYMMETRY

For a quantum mechanical system of N particles, the Hamiltonian is a sum of the total kinetic energy and the potential energy

$$H = \sum_i \frac{p_i^2}{2\mu} + V(\mathbf{r}_1, \cdots, \mathbf{r}_N). \tag{7.1}$$

In the presence of an external field, the potential can be also a function of time.

In classical physics, it is possible to keep track of individual particles. For example, when we have a system of two particles, we can follow the trajectory of these two particles separately at each instant of time. However, in quantum mechanics, identical particles are indistinguishable and we cannot follow their trajectories.

Consider, for example, a N-particle system. The wave function can be generally expressed as

$$|\Psi\rangle = |n_1 \cdots n_i \cdots n_j \cdots n_N\rangle, \tag{7.2}$$

where $|n_i\rangle$ is the state of the i'th particle taking. Since the system consists of identical particles, this state should remain the same if any two particles, say i and j are exchanged. This operation is carried out by the permutation operator P_{ij}

$$P_{ij} |n_1 \cdots n_i \cdots n_j \cdots n_N\rangle = |n_1 \cdots n_j \cdots n_i \cdots n_N\rangle. \tag{7.3}$$

Under the transformation P_{ij}, the states n_i and n_j are swapped. Since all the particles are identical, the states before and after the permutation are unchanged, hence $|n_1 \cdots n_i \cdots n_j \cdots n_N\rangle$ should be an eigenstate of P_{ij}

$$P_{ij} |n_1 \cdots n_i \cdots n_j \cdots n_N\rangle = \lambda_{ij} |n_1 \cdots n_i \cdots n_j \cdots n_N\rangle, \tag{7.4}$$

where λ_{ij} is arbitrary constant. Since $P_{ij} = P_{ji}$ and a second exchange of these two particles recreates the original state, we have

$$P_{ij}^2 |\Psi\rangle = \lambda_{ij}^2 |\Psi\rangle = |\Psi\rangle. \tag{7.5}$$

Thus λ_{ij} can take two values

$$\lambda_{ij} = \pm 1. \tag{7.6}$$

As all particles are identical, the exchange of particles should act in the same way on the wave function. Hence λ_{ij} should take the same value, either 1 or -1, for any pair of particles. This means that the wave function should either change sign or remain unchanged upon the exchange of two particles. Hence either

$$P_{ij} |\Psi\rangle = |\Psi\rangle, \quad \forall (i, j; i \neq j), \tag{7.7}$$

DOI: 10.1201/9781003174882-7

or

$$P_{ij}|\Psi\rangle = -|\Psi\rangle, \quad \forall (i, j; i \neq j), \tag{7.8}$$

holds. The wave function that satisfies Eq. (7.7) with the eigenvalue $+1$ is called symmetric and that satisfies Eq. (7.8) with the eigenvalue -1 is called antisymmetric, with respect to the exchange of particles.

Furthermore, as the interaction between identical particles is always symmetric under their exchange

$$V(\mathbf{r}_1 \cdots \mathbf{r}_i \cdots \mathbf{r}_j \cdots \mathbf{r}_N) = V(\mathbf{r}_1 \cdots \mathbf{r}_j \cdots \mathbf{r}_i \cdots \mathbf{r}_N), \tag{7.9}$$

the matrix elements of this interaction between symmetric and anti-symmetric states should vanish

$$\langle \Psi_s | V(\mathbf{r}_1 \cdots \mathbf{r}_i \cdots \mathbf{r}_j \cdots \mathbf{r}_N) | \Psi_a \rangle = 0, \tag{7.10}$$

where $|\Psi_s\rangle$ is a symmetric state and $|\Psi_a\rangle$ an anti-symmetric state. Therefore, no transitions take place between the two states with different symmetries under the exchange of particles.

7.2   BOSE-EINSTEIN AND FERMI-DIRAC STATISTICS

A system of N identical particles is said to satisfy Bose-Einstein statistics if its wave function is symmetric under the exchange of particles, and these particles are known as bosons (named after S. N. Bose). On the other hand, a system of N identical particles is said to satisfy Fermi-Dirac statistics if its wave function is antisymmetric under the exchange of particles, and these particles are known as fermions (named after E. Fermi).

These two kinds of particles are distinguished by their spins: half-integer-spin particles are fermions and integer-spin particles are bosons. This spin-statistics connection is a law of nature. In the nonrelativistic quantum mechanics, this principle must be accepted as an empirical postulate. In the relativistic quantum theory, it can be proven that half-integer-spin particles cannot be bosons and integer-spin particles cannot be fermions.

The antisymmetry of the fermion wave function is equivalent to Pauli's exclusion principle, empirically formulated by Pauli in 1925. It states that *two fermions cannot occupy the same individual state.*

To understand this, let us assume two fermions, say fermion i and fermion j, are in the same quantum state, i.e. $n_i = n_j$. Then the antisymmetry requires

$$|n_1 \cdots n_i \cdots n_i \cdots n_N\rangle = -|n_1 \cdots n_i \cdots n_i \cdots n_N\rangle = 0. \tag{7.11}$$

7.2.1   EXCHANGE DEGENERACY

We consider a system of N identical particles without interactions, where the total Hamiltonian is simply a sum of the Hamiltonian for each particle

$$H = H_1(\mathbf{r}_1) + H_2(\mathbf{r}_2) + \cdots + H_N(\mathbf{r}_N), \tag{7.12}$$

where H_i is the Hamiltonian of the i'th particle. $H_i(r_i)$ is particle distinguishable because they act on different particles. The corresponding Schrödinger equation is

$$H\Psi(\mathbf{r}_1,\cdots,\mathbf{r}_N) = E\Psi(\mathbf{r}_1,\cdots,\mathbf{r}_N). \tag{7.13}$$

If we use $\varphi_n(\mathbf{r}_i)$ to represent the n'th eigenfunction of the i'th particle and E_n the corresponding eigenenergy, namely

$$H_i(\mathbf{r}_i)\varphi_n(\mathbf{r}_i) = E_n\varphi_n(\mathbf{r}_i). \tag{7.14}$$

Clearly the product of N single-particle states is a solution of the above Schrödinger equation

$$\Psi(\mathbf{r}_1,\cdots,\mathbf{r}_N) = \varphi_{n_1}(\mathbf{r}_1)\cdots\varphi_{n_i}(\mathbf{r}_i)\cdots\varphi_{n_N}(\mathbf{r}_N). \tag{7.15}$$

More general, we denote this state as

$$|\Psi\rangle = |\varphi_{n_1}^1\rangle|\varphi_{n_2}^2\rangle\cdots|\varphi_{n_i}^i\rangle\cdots|\varphi_{n_N}^N\rangle. \tag{7.16}$$

Here we use the superscript to denote the index of particles: $|\varphi_n^i\rangle$ is the quantum state of the i'th particle and its wave function is $\psi_n(\mathbf{r}_i) = \langle\mathbf{r}_i|\varphi_n^i\rangle$. The corresponding hermitian conjugate of $|\Psi\rangle$ is

$$\langle\Psi| = \langle\varphi_{n_N}^N|\cdots\langle\varphi_{n_i}^i|\cdots\langle\varphi_{n_2}^2|\langle\varphi_{n_1}^1|,$$

we order the particles from right to left, namely the quantum state of the first particle, $\langle n_1|$, is at the rightmost position, then the states from right to left are those of the second, third, and other particles, respectively. This is the notation we used below.

The above state Ψ has an energy

$$E = \sum_i E_{n_i}. \tag{7.17}$$

Because of the indistinguishability of the particles, we are not able to say which particle is in which state. This means that there are many different combinations of single-particle wave functions that give the same energy. This degeneracy in the eigenenergy is called exchange degeneracy.

However, $|\Psi\rangle$ above defined is generally not symmetric, neither antisymmetric. The requirement of wave function being symmetric or antisymmetric would completely eliminate this degeneracy. In fact, the entire space of functions spanned by the eigenfunctions contains only one symmetric and one anti-symmetric wave function.

7.2.2 ANTI-SYMMETRIZED WAVE FUNCTIONS

For fermions, the wave function after antisymmetrization is a Slater's determinant of an $N \times N$ matrix

$$\Psi(\mathbf{r}_1,\cdots,\mathbf{r}_N) = \frac{1}{\sqrt{N!}}\begin{vmatrix} \varphi_{n_1}(\mathbf{r}_1) & \varphi_{n_2}(\mathbf{r}_1) & \cdots & \varphi_{n_N}(\mathbf{r}_1) \\ \varphi_{n_1}(\mathbf{r}_2) & \varphi_{n_2}(\mathbf{r}_2) & \cdots & \varphi_{n_N}(\mathbf{r}_2) \\ \vdots & \vdots & \ddots & \vdots \\ \varphi_{n_1}(\mathbf{r}_N) & \varphi_{n_2}(\mathbf{r}_N) & \cdots & \varphi_{n_N}(\mathbf{r}_N) \end{vmatrix}. \tag{7.18}$$

This determinant can be expressed using the Leibniz formula as

$$\Psi(\mathbf{r}_1,\cdots,\mathbf{r}_N) = \frac{1}{\sqrt{N!}} \sum_{\sigma \in S_N} \text{sgn}(\sigma)\, \varphi_{k_1}(\mathbf{r}_{\sigma_1})\, \varphi_{k_2}(\mathbf{r}_{\sigma_2}) \cdots \varphi_{k_N}(\mathbf{r}_{\sigma_N}), \qquad (7.19)$$

where the sum is over all permutations σ of $(1,2,\ldots,N)$. A permutation is to reorder this set of integers. The i'th position becomes σ_i after the reordering. The whole set of all such permutations is denoted by S_N. For each permutation, $\text{sgn}(\sigma)$ equals 1 if the reordering given by σ can be achieved by successively interchanging two entries an even number of times starting from the natural ordering $(1,2\cdots,N)$, and -1 if it can be achieved by an odd number of such interchanges.

7.2.3 SYMMETRIZED WAVE FUNCTIONS

For bosons, the wave function after symmetrization can be formally expressed as

$$\Psi(\mathbf{r}_1,\cdots,\mathbf{r}_N) = C \begin{vmatrix} \varphi_{n_1}(\mathbf{r}_1) & \varphi_{n_2}(\mathbf{r}_1) & \cdots & \varphi_{n_N}(\mathbf{r}_1) \\ \varphi_{n_1}(\mathbf{r}_2) & \varphi_{n_2}(\mathbf{r}_2) & \cdots & \varphi_{n_N}(\mathbf{r}_2) \\ \vdots & \vdots & \ddots & \vdots \\ \varphi_{n_1}(\mathbf{r}_N) & \varphi_{n_2}(\mathbf{r}_N) & \cdots & \varphi_{n_N}(\mathbf{r}_N) \end{vmatrix}_+ . \qquad (7.20)$$

In the expression $|*|_+$ is not a determinant. It is defined by taking the sign function $\text{sgn}(\sigma) = 1$ for all permutations σ, so that

$$\Psi(r_1,\cdots,r_N) = C \sum_{\sigma \in S_N} \varphi_{k_1}(r_{\sigma_1})\, \varphi_{k_2}(r_{\sigma_2}) \cdots \varphi_{k_N}(r_{\sigma_N}), \qquad (7.21)$$

Since two bosons can be in the same states, the normalization constant C depends on the quantum states occuiped, i.e. $(n_1 \cdots n_N)$. If none of $(n_1 \cdots n_N)$ are equal to each other, $C = 1/\sqrt{N!}$.

7.2.4 TWO IDENTICAL PARTICLES

As an example, let us consider a system with only two particles. We assume one of the particles is in the state $|n\rangle$ and the other is in the state $|m\rangle$. If these two particles are distinguishable, their wave function is simply given by

$$|\Psi_c\rangle = |\varphi_n^1\rangle |\varphi_m^2\rangle. \qquad (7.22)$$

However, if the two particles are not distinguishable, the wave function should be either symmetrized

$$\boxed{|\Psi_s\rangle = C_s \left(|\varphi_n^1\rangle |\varphi_m^2\rangle + |\varphi_m^1\rangle |\varphi_n^2\rangle \right),} \qquad (7.23)$$

if they are bosons or antisymmetrized

$$\boxed{|\Psi_a\rangle = C_a \left(|\varphi_n^1\rangle |\varphi_m^2\rangle - |\varphi_m^1\rangle |\varphi_n^2\rangle \right),} \qquad (7.24)$$

if they are two fermions. C_s and C_a are normalization constants. If $m \neq n$, it is simple to show that $C_s = C_a = 1/\sqrt{2}$.

On the other hand, if $m = n$, the symmetrized state becomes

$$|\Psi_s\rangle = |\varphi_n^1\rangle |\varphi_n^2\rangle. \tag{7.25}$$

However, the antisymmetrized state $|\Psi_a\rangle$ becomes zero, as a consequence of Pauli exclusion principle.

7.2.5 EXCHANGE FORCE

To understand how the symmetrization would affect the behavior of two particles, let us evaluate the expectation value of the square of the distance between the two particles

$$\Delta = \left\langle (\mathbf{r}_1 - \mathbf{r}_2)^2 \right\rangle. \tag{7.26}$$

If the two particles are distinguishable, then

$$
\begin{aligned}
\Delta_c &= \left\langle \Psi_c \left| (\mathbf{r}_1 - \mathbf{r}_2)^2 \right| \Psi_c \right\rangle \\
&= \left\langle \varphi_m^2 \right| \left\langle \varphi_n^1 \right| (\mathbf{r}_1^2 + \mathbf{r}_2^2 - 2\mathbf{r}_1 \cdot \mathbf{r}_2) \left| \varphi_n^1 \right\rangle \left| \varphi_m^2 \right\rangle \\
&= \left\langle \varphi_n^1 \left| \mathbf{r}_1^2 \right| \varphi_n^1 \right\rangle + \left\langle \varphi_m^2 \left| \mathbf{r}_2^2 \right| \varphi_m^2 \right\rangle - 2\left\langle \varphi_n^1 \left| \mathbf{r}_1 \right| \varphi_n^1 \right\rangle \cdot \left\langle \varphi_m^2 \left| \mathbf{r}_2 \right| \varphi_m^2 \right\rangle \\
&= \left\langle r^2 \right\rangle_n + \left\langle r^2 \right\rangle_m - 2 \left\langle r \right\rangle_n \cdot \left\langle r \right\rangle_m
\end{aligned}
\tag{7.27}
$$

where

$$\langle \mathbf{r} \rangle_\alpha = \int \mathbf{r} |\psi_\alpha(\mathbf{r})|^2 \, d\mathbf{r}, \tag{7.28}$$

$$\langle r^2 \rangle_\alpha = \int r^2 |\psi_\alpha(\mathbf{r})|^2 \, d\mathbf{r}, \tag{7.29}$$

and $\alpha = m$ or n.

For the anti-symmetrized wave function, the expectation value of the square distance is

$$\Delta_a = \left\langle \Psi_a \left| (\mathbf{r}_1 - \mathbf{r}_2)^2 \right| \Psi_a \right\rangle = \left\langle \Psi_a \left| \mathbf{r}_1^2 + \mathbf{r}_2^2 - 2\mathbf{r}_1 \cdot \mathbf{r}_2 \right| \Psi_a \right\rangle. \tag{7.30}$$

For fermions, $m \neq n$, the average of $\mathbf{r}_1^2$ is

$$
\begin{aligned}
\left\langle \Psi_a \left| \mathbf{r}_1^2 \right| \Psi_a \right\rangle &= \frac{1}{2} \left(\left(\langle \varphi_m^2 | \langle \varphi_n^1 | - \langle \varphi_n^2 | \langle \varphi_m^1 | \right) \mathbf{r}_1^2 \left(|\varphi_n^1\rangle |\varphi_m^2\rangle - |\varphi_m^1\rangle |\varphi_n^2\rangle \right) \right) \\
&= \frac{1}{2} \left\langle \varphi_n^1 \left| \mathbf{r}_1^2 \right| \varphi_n^1 \right\rangle + \frac{1}{2} \left\langle \varphi_m^1 \left| \mathbf{r}_1^2 \right| \varphi_m^1 \right\rangle \\
&= \frac{1}{2} \left(\langle r^2 \rangle_n + \langle r^2 \rangle_m \right).
\end{aligned}
\tag{7.31}
$$

Similarly, we have

$$\left\langle \Psi_a \left| \mathbf{r}_2^2 \right| \Psi_a \right\rangle = \frac{1}{2} \left(\langle r^2 \rangle_n + \langle r^2 \rangle_m \right). \tag{7.32}$$

$\langle \Psi_a | \mathbf{r}_2^2 | \Psi_a \rangle$ equals $\langle \Psi_a | \mathbf{r}_1^2 | \Psi_a \rangle$ because the particles are indistinguishable. The average of $2\mathbf{r}_1 \cdot \mathbf{r}_2$ is

$$
\begin{aligned}
\langle \Psi_a | 2\mathbf{r}_1 \cdot \mathbf{r}_2 | \Psi_a \rangle &= \left(\left(\langle \varphi_m^2 | \langle \varphi_n^1 | - \langle \varphi_n^2 | \langle \varphi_m^1 | \right) (\mathbf{r}_1 \cdot \mathbf{r}_2) \left(|\varphi_n^1\rangle |\varphi_m^2\rangle - |\varphi_m^1\rangle |\varphi_n^2\rangle \right) \right) \\
&= \langle \varphi_n^1 | \mathbf{r}_1 | \varphi_n^1 \rangle \cdot \langle \varphi_m^2 | \mathbf{r}_2 | \varphi_m^2 \rangle + \langle \varphi_m^1 | \mathbf{r}_1 | \varphi_m^1 \rangle \cdot \langle \varphi_n^2 | \mathbf{r}_2 | \varphi_n^2 \rangle \\
&\quad - \langle \varphi_n^1 | \mathbf{r}_1 | \varphi_m^1 \rangle \cdot \langle \varphi_m^2 | \mathbf{r}_2 | \varphi_n^2 \rangle + \langle \varphi_m^1 | \mathbf{r}_1 | \varphi_n^1 \rangle \cdot \langle \varphi_n^2 | \mathbf{r}_2 | \varphi_m^2 \rangle \\
&= 2\langle \mathbf{r} \rangle_n \cdot \langle \mathbf{r} \rangle_m - 2\langle \mathbf{r} \rangle_{nm} \cdot \langle \mathbf{r} \rangle_{mn} \\
&= 2\langle \mathbf{r} \rangle_n \cdot \langle \mathbf{r} \rangle_m - 2|\langle \mathbf{r} \rangle_{nm}|^2 , \quad\quad (7.33)
\end{aligned}
$$

where

$$
\langle \mathbf{r} \rangle_{nm} = \int \mathbf{r} \psi_n^* (\mathbf{r}) \, \psi_m (\mathbf{r}) \, d\mathbf{r} = \langle \mathbf{r} \rangle_{mn}^* . \quad\quad (7.34)
$$

Thus we have

$$
\boxed{\Delta_a = \langle r^2 \rangle_n + \langle r^2 \rangle_m - 2\langle \mathbf{r} \rangle_n \cdot \langle \mathbf{r} \rangle_m + 2|\langle \mathbf{r} \rangle_{nm}|^2 = \Delta_c + 2|\langle \mathbf{r} \rangle_{nm}|^2 .} \quad\quad (7.35)
$$

For the symmetrized wave function, the expectation value of the square distance can be similarly calculated. If $m \neq n$, it is simple to show that

$$
\boxed{\Delta_s = \left\langle \Psi_s \left| (\mathbf{r}_1 - \mathbf{r}_2)^2 \right| \Psi_s \right\rangle = \Delta_c - 2|\langle \mathbf{r} \rangle_{nm}|^2 .} \quad\quad (7.36)
$$

If $m = n$, however,

$$
\boxed{\Delta_s = \left\langle \Psi_s \left| (\mathbf{r}_1 - \mathbf{r}_2)^2 \right| \Psi_s \right\rangle = \Delta_c .} \quad\quad (7.37)
$$

The above results indicate that the distance between the two particles in a spatial symmetrized state is smaller than its classical counterpart if $m \neq n$ or equal to its classical counterpart if $m = n$, while that between the two particles in a spatial anti-symmetrized state is always larger than its classical counterpart. Thus the system behaves as though there were an attractive force between two symmetrized particles, pulling them closer together, and a repulsive force between two anti-symmetrized particles, pushing them apart. This putative force is called exchange force, which is a consequence of the symmetrization and a strictly quantum mechanical phenomenon, with no classical counterpart. However, this exchange force disappears if two bosons are in the same states.

The above discussion on the attractive or repulsive nature of the exchange force is based on the symmetry of the spatial wave function. If a particle possesses spin or other internal degrees of freedom, the complete state of the whole system includes not only its spatial wave function, but also the wave function describing the internal degrees of freedom. In this case, even fermions can be in a spatially symmetric state whose exchange force is attractive if the internal degree of freedom is in an antisymmetric state. Similarly, a boson system can be in a spatially anti-symmetric state with a repulsive exchange force if its internal degree of freedom is also in an anti-symmetric state.

7.3 FREE FERMION GAS

The Pauli exclusion principle plays a critical role in programming physical properties of electrons or more generally fermions. To understand this, let us consider a system of noninteracting fermions by ignoring all forces except the "statistical force" resulted from the antisymmetrization of the wave fucntion.

7.3.1 PARTICLE IN A PERIODIC BOX

We first consider a free particle in a one-dimensional system of length L with periodic boundary condition. The eigen-equation is simply given by

$$-\frac{\hbar^2}{2\mu}\frac{\partial^2}{\partial x^2}\psi(x) = \varepsilon\psi(x). \tag{7.38}$$

The solution of the eigenfunction that satisfies the periodic boundary condition $\psi(x+L) = \psi(x)$ is simply a plane wave

$$\psi_k(x) = \frac{1}{\sqrt{L}}e^{ikx}, \tag{7.39}$$

where k is the wave vector

$$k = \frac{2\pi n}{L} \tag{7.40}$$

and n is an integer. Here we add a subscript k to label each eigenstate. The corresponding eigenenergy

$$\varepsilon_k = \frac{\hbar^2 k^2}{2\mu} \tag{7.41}$$

is a parabolic function of k.

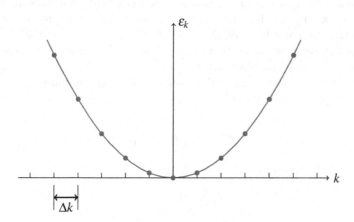

In the k-space, the eigen levels are evenly distributed. Each state occupies a k-interval

$$\Delta k = \frac{2\pi}{L}. \tag{7.42}$$

Now we generalize the above results to three dimensions, and consider a free fermion in a box of dimensions $L \times L \times L$ with periodic boundary conditions. The eigen-equation is

$$-\frac{\hbar^2}{2\mu}\nabla^2\Psi(x,y,z) = E\Psi(x,y,z). \tag{7.43}$$

It is simple to show that the eigenfunction $\Psi(x,y,z)$ is a product of three eigenfunctions along the three coordinate axes

$$\Psi_{\mathbf{k}}(x,y,z) = \psi_{k_x}(x)\psi_{k_y}(y)\psi_{k_z}(z) = \frac{1}{L^{3/2}}e^{ik_x x + ik_y y + ik_z z}, \tag{7.44}$$

where

$$\mathbf{k} = (k_x, k_y, k_z) = \frac{2\pi}{L}(n_x, n_y, n_z) \tag{7.45}$$

is the wave vector. The corresponding eigenenergy is

$$E_{\mathbf{k}} = \varepsilon_{k_x} + \varepsilon_{k_y} + \varepsilon_{k_z} = \frac{\hbar^2 k^2}{2\mu}. \tag{7.46}$$

Again the eigenstates are uniformly distributed in $\mathbf{k}$-space, and each state occupies a volume

$$\Delta\mathbf{k} = \left(\frac{2\pi}{L}\right)^3 = \frac{8\pi^3}{L^3}. \tag{7.47}$$

7.3.2 FERMI SURFACE

We now put N fermions into the box (for an object of macroscopic size, N is of the order of Avogadro's number, which is of order 6×10^{23}). If the internal degrees of freedom of fermions, such as spins, are not considered, then no more than one fermion can occupy one eigenstate due to the Pauli exclusion principle. In the ground state, there fermions will fill up the lowest N-eigenenergy states. The filled wave vectors will form a sphere in the $\mathbf{k}$-space in the large-N limit. The radius of this sphere is called the Fermi wave vector k_F, which is determined by the fact that each fermion requires a valume $\Delta\mathbf{k}$ and the volume of the sphere must equal the total volume occupied by fermions, i.e.

$$\frac{4}{3}\pi k_F^3 = \frac{8\pi^3 N}{L^3} = \frac{8\pi^3 N}{V}, \tag{7.48}$$

where $V = L^3$ is the volume of the box. Thus

$$k_F = \left(6\pi^2\rho\right)^{1/3}, \tag{7.49}$$

where

$$\rho = \frac{N}{V} \tag{7.50}$$

is the fermion density, namely the number of fermions per unit volume.

The surface of the sphere spanned by $|\mathbf{k}| = k_F$ separates the occupied and unoccupied states. It is called the Fermi surface. The corresponding energy is called Fermi energy or Fermi level, denoted as E_F. For the fermion gas considered here

$$E_F = \frac{\hbar^2}{2\mu} \left(6\pi^2 \rho\right)^{2/3}. \tag{7.51}$$

7.3.3 DEGENERACY PRESSURE

The energy of the whole system in the ground state is a sum of the energies of all occupied states. It is given by integrating over all the energies up to the Fermi level.

$$
\begin{aligned}
E_0 &= \sum_{|\mathbf{k}| \le k_F} \frac{\hbar^2 \mathbf{k}^2}{2\mu} \\
&= \frac{V}{(2\pi)^3} \sum_{|\mathbf{k}| \le k_F} \frac{\hbar^2 \mathbf{k}^2}{2\mu} \Delta \mathbf{k} \\
&= \frac{V}{(2\pi)^3} \int_{|\mathbf{k}| \le k_F} d\mathbf{k} \frac{\hbar^2 \mathbf{k}^2}{2\mu} \\
&= \frac{V}{(2\pi)^3} \int_0^{k_F} 4\pi k^2 dk \frac{\hbar^2 k^2}{2\mu} \\
&= \frac{V \hbar^2 k_F^5}{20\pi^2 \mu}.
\end{aligned}
\tag{7.52}
$$

In terms of the fermion density

$$E_0 = \frac{V\hbar^2}{20\pi^2\mu} \left(6\pi^2\rho\right)^{5/3} = \frac{3\hbar^2}{10\mu V^{2/3}} \left(6\pi^2\right)^{2/3} N^{5/3}. \tag{7.53}$$

This energy exerts a pressure to the box. If the box expands, the energy decreases. The change of the total energy per volume change is the pressure applied to the box

$$P = \frac{dE_0}{dV} = -\frac{2}{3V} \frac{3\hbar^2}{10\mu V^{2/3}} \left(6\pi^2\right)^{2/3} N^{5/3} = -\frac{2E_0}{3V}. \tag{7.54}$$

This is an internal pressure induced by the Pauli exclusion principle. It is called degeneracy pressure.

7.4 HYDROGEN MOLECULE

A hydrogen molecule contains two protons and two electrons. Given the positions of two protons, the Hamiltonian for the two electrons is

$$H = \frac{1}{2\mu} \left(\mathbf{p}_1^2 + \mathbf{p}_2^2\right) + \frac{e^2}{4\pi\varepsilon_0} \left(\frac{1}{r_{12}} - \frac{1}{r_{1a}} - \frac{1}{r_{2a}} - \frac{1}{r_{1b}} - \frac{1}{r_{2b}}\right), \tag{7.55}$$

where μ is the mass of electron, $r_{12} = |\mathbf{r}_1 - \mathbf{r}_2|$, $\mathbf{r}_1$ and $\mathbf{r}_2$ are the coordinates of two electrons. $\mathbf{r}_{ia} = \mathbf{r}_i - \mathbf{r}_a$ and $\mathbf{r}_{ib} = \mathbf{r}_i - \mathbf{r}_b$ with $\mathbf{r}_a$ and $\mathbf{r}_b$ the coordinates of two protons. Electrons have spins. This Hamiltonian does not depend explicitly on the spins of electrons. But the wave function depends on spins.

If we use σ_1 and σ_2 to denote their spins, the eigen-equation of these two electrons can be written as

$$H\Psi(\mathbf{r}_1\sigma_1, \mathbf{r}_2\sigma_2) = E\Psi(\mathbf{r}_1\sigma_1, \mathbf{r}_2\sigma_2). \tag{7.56}$$

As the Hamiltonian does not depend on σ_1 and σ_2, the coordinate and spin variables are separated and the eigenfunction can be written as a product of the spatial wave function $\psi(\mathbf{r}_1, \mathbf{r}_2)$ and the spin wave function $\chi(\sigma_1, \sigma_2)$

$$\Psi(\mathbf{r}_1\sigma_1, \mathbf{r}_2\sigma_2) = \psi(\mathbf{r}_1, \mathbf{r}_2)\chi(\sigma_1, \sigma_2). \tag{7.57}$$

As $\Psi(\mathbf{r}_1\sigma_1, \mathbf{r}_2\sigma_2)$ is antisymmetric under the exchange of the two electrons, two cases emerge:

1. $\psi(\mathbf{r}_1, \mathbf{r}_2)$ is symmetric under the exchange of $\mathbf{r}_1$ and $\mathbf{r}_2$, and $\chi(\sigma_1, \sigma_2)$ is antisymmetric under the exchange of σ_1 and σ_2

$$\Psi_{\text{singlet}}(\mathbf{r}_1\sigma_1, \mathbf{r}_2\sigma_2) = \psi_s(\mathbf{r}_1, \mathbf{r}_2)\chi_a(\sigma_1, \sigma_2), \tag{7.58}$$

   where the subscripts s and a represent symmetric and antisymmetric, respectively. For a system of two $S = 1/2$ spins, the anti-symmetric function is unique and given by

$$\chi_a(\sigma_1, \sigma_2) = \frac{1}{\sqrt{2}}(\delta_{\sigma_1,1}\delta_{\sigma_2,-1} - \delta_{\sigma_1,-1}\delta_{\sigma_2,1}). \tag{7.59}$$

   This is a state of total spin 0 or a spin singlet. The corresponding total wave function is denoted by Ψ_{singlet}.

2. $\psi(\mathbf{r}_1, \mathbf{r}_2)$ is antisymmetric under the exchange of $\mathbf{r}_1$ and $\mathbf{r}_2$, and $\chi(\sigma_1, \sigma_2)$ is symmetric under the exchange of σ_1 and σ_2

$$\Psi_{\text{triplet}}(\mathbf{r}_1\sigma_1, \mathbf{r}_2\sigma_2) = \psi_a(\mathbf{r}_1, \mathbf{r}_2)\chi_s(\sigma_1, \sigma_2). \tag{7.60}$$

   The symmetric spin wave function χ_s is an eigenstate of total spin 1. It is three-fold degenerate and can be represented as

$$\chi_s(\sigma_1, \sigma_2) = \begin{cases} \delta_{\sigma_1,1}\delta_{\sigma_2,1}, & m = 1, \\ \frac{1}{\sqrt{2}}(\delta_{\sigma_1,1}\delta_{\sigma_2,-1} + \delta_{\sigma_1,-1}\delta_{\sigma_2,1}), & m = 0, \\ \delta_{\sigma_1,-1}\delta_{\sigma_2,-1}, & m = -1, \end{cases} \tag{7.61}$$

where m is the third component value of the total spin.

The eigen-equation is difficult to solve. To the leading order approximation, we use the ground state wave function of a hydrogen atom, ψ_0, to construct two trial functions ψ_s and ψ_a as

$$\psi_s(\mathbf{r}_1, \mathbf{r}_2) = C_s[\psi_0(\mathbf{r}_{1a})\psi_0(\mathbf{r}_{2b}) + \psi_0(\mathbf{r}_{2a})\psi_0(\mathbf{r}_{1b})], \qquad (7.62)$$

$$\psi_a(\mathbf{r}_1, \mathbf{r}_2) = C_a[\psi_0(\mathbf{r}_{1a})\psi_0(\mathbf{r}_{2b}) - \psi_0(\mathbf{r}_{2a})\psi_0(\mathbf{r}_{1b})], \qquad (7.63)$$

where C_s and C_a are normalization constants. C_s is determined by the equation

$$
\begin{aligned}
1 &= \int d\mathbf{r}_1 d\mathbf{r}_2 |\psi_s(\mathbf{r}_1, \mathbf{r}_2)|^2 \\
&= C_s^2 \int d\mathbf{r}_1 d\mathbf{r}_2 [\psi_0(\mathbf{r}_{1a})\psi_0(\mathbf{r}_{2b}) + \psi_0(\mathbf{r}_{2a})\psi_0(\mathbf{r}_{1b})]^2 \\
&= C_s^2 \int d\mathbf{r}_1 d\mathbf{r}_2 [\psi_0^2(\mathbf{r}_{1a})\psi_0^2(\mathbf{r}_{2b}) + \psi_0^2(\mathbf{r}_{2a})\psi_0^2(\mathbf{r}_{1b}) \\
&\qquad\qquad + 2\psi_0(\mathbf{r}_{1a})\psi_0(\mathbf{r}_{2b})\psi_0(\mathbf{r}_{2a})\psi_0(\mathbf{r}_{1b})] \\
&= C_s^2(2 + 2A), \qquad (7.64)
\end{aligned}
$$

where

$$A = \int d\mathbf{r}_1 d\mathbf{r}_2 \psi_0^*(\mathbf{r}_{1a}) \psi_0^*(\mathbf{r}_{2b}) \psi_0(\mathbf{r}_{2a}) \psi_0(\mathbf{r}_{1b}) \qquad (7.65)$$

is the exchange overlap between the ground state wave functions of two electrons. Similarly, C_a is determined by the equation

$$1 = \int d\mathbf{r}_1 d\mathbf{r}_2 |\psi_a(\mathbf{r}_1, \mathbf{r}_2)|^2 = C_a^2(2 - A). \qquad (7.66)$$

We thus have

$$C_s = \frac{1}{\sqrt{2 + 2A}}, \qquad (7.67)$$

$$C_a = \frac{1}{\sqrt{2 - 2A}}. \qquad (7.68)$$

When the separation between the protons is large, A is small. For simplicity in the discussion below, we neglect A and take $C_s = C_a = 1/\sqrt{2}$. In this case

$$\langle \Psi_{\text{singlet}} | H | \Psi_{\text{singlet}} \rangle = \langle \psi_s | H | \psi_s \rangle = 2E_0 + V + J, \qquad (7.69)$$

$$\langle \Psi_{\text{triplet}} | H | \Psi_{\text{triplet}} \rangle = \langle \psi_a | H | \psi_a \rangle = 2E_0 + V - J, \qquad (7.70)$$

where E_0 is the ground state energy of hydrogen atom

$$\left(-\frac{\hbar^2}{2\mu}\nabla^2 - \frac{e^2}{4\pi\varepsilon_0 r} \right) \psi_0(\mathbf{r}) = E_0 \psi_0(\mathbf{r}). \qquad (7.71)$$

V is the Coulomb potential of an electron exerted by the second electron and the second proton

$$
\begin{aligned}
V &= \frac{e^2}{4\pi\varepsilon_0} \int d\mathbf{r}_1 d\mathbf{r}_2 \left(\frac{1}{r_{12}} - \frac{1}{r_{2a}} - \frac{1}{r_{1b}} \right) \psi_0^2(\mathbf{r}_{1a}) \psi_0^2(\mathbf{r}_{2b}) \\
&= \frac{e^2}{4\pi\varepsilon_0} \int d\mathbf{r}_1 d\mathbf{r}_2 \psi_0^2(\mathbf{r}_{1a}) \frac{1}{r_{12}} \psi_0^2(\mathbf{r}_{2b}) - \frac{e^2}{2\pi\varepsilon_0} \int d\mathbf{r}_1 \frac{1}{r_{1b}} \psi_0^2(\mathbf{r}_{1a}). \quad (7.72)
\end{aligned}
$$

J is the exchange energy, without a classical correspondence

$$J = \frac{e^2}{4\pi\varepsilon_0} \int d\mathbf{r}_1 d\mathbf{r}_2 \, \psi_0(\mathbf{r}_{2a}) \, \psi_0(\mathbf{r}_{1b}) \left(\frac{1}{r_{12}} - \frac{1}{r_{2a}} - \frac{1}{r_{1b}} \right) \psi_0(\mathbf{r}_{1a}) \, \psi_0(\mathbf{r}_{2b}). \quad (7.73)$$

Both V and J depend on the distance between the two protons. Given ψ_0, it tunes out that the exchange energy is always negative, i.e. $J < 0$. Therefore, the energy of the singlet is lower than the triplet

$$\left\langle \Psi_{\text{singlet}} \middle| H \middle| \Psi_{\text{singlet}} \right\rangle < \left\langle \Psi_{\text{triplet}} \middle| H \middle| \Psi_{\text{triplet}} \right\rangle. \quad (7.74)$$

Hence the symmetric wave function has a lower energy. It indicates that the symmetric spatial wave function leads to a bonding configuration of electrons and the antisymmetric one does not. The actual electron charge density is given by the square of the magnitude of the wave function, and the symmetric wave function gives a high electron density between two protons, leading to a net attractive force, i.e. a covalent bond, between two hydrogen atoms.

In a hydrogen molecule, two hydrogen atoms share two electrons via covalent bonding, which is purely a quantum effect originated from the exchange interaction of Fermi-Dirac statistics of electrons. Apparently, spin plays an important role in the formation of covalence bond in a hydrogen molecule. It is the antisymmetry of the spin configuration with respect to exchange (i.e. spin singlet) that enables the spatial wave function of the two electrons to be symmetric, so that the exchange force is attractive. Thus forming a spin singlet covalence bond can lower the energy of the molecule. On the other hand, if the configuration of spins is symmetric, namely in a spin triplet state, the spatial configuration of electrons has to be antisymmetric with respect to exchange, leading to a repulsive exchange interaction. Thus a pair of electrons in a spin triplet state is chemically unstable.

The energy difference between the triplet and the singlet states is determined purely by the exchange energy

$$\left\langle \Psi_{\text{triplet}} \middle| H \middle| \Psi_{\text{triplet}} \right\rangle - \left\langle \Psi_{\text{singlet}} \middle| H \middle| \Psi_{\text{singlet}} \right\rangle = -2J. \quad (7.75)$$

It implies that the low-energy physics of hydrogen molecules are governed by the spin states of the two electrons, while the spatial wave function is just to change the exchange constant J. It further suggests that we can model this singlet and triplet system by the interaction between the spin operators of the two electrons, $\mathbf{S}_1 = \hbar\boldsymbol{\sigma}_1/2$ and $\mathbf{S}_2 = \hbar\boldsymbol{\sigma}_2/2$. Since the system is invariant under the spin rotation, the interaction should be determined by the scaler product of two spin operators, i.e. $\mathbf{S}_1 \cdot \mathbf{S}_2$. This scalar product is related to the total spin operator $\mathbf{S}_1 + \mathbf{S}_2$ by the relation

$$\mathbf{S}_1 \cdot \mathbf{S}_2 = \frac{1}{2} \left[(\mathbf{S}_1 + \mathbf{S}_2)^2 - \mathbf{S}_1^2 - \mathbf{S}_2^2 \right] = \frac{1}{2} (\mathbf{S}_1 + \mathbf{S}_2)^2 - \frac{3}{4}\hbar^2. \quad (7.76)$$

It is simple to show that $\mathbf{S}_1 \cdot \mathbf{S}_2$ equals $-3\hbar^2/4$ and $\hbar^2/4$ for the spin singlet and triplet states, respectively. Thus the low energy states of hydrogen molecules can be mimicked by the Hamiltonian,

$$H_{\text{ex}} = -\frac{2J}{\hbar^2}\left(\mathbf{S}_1 \cdot \mathbf{S}_2 + \frac{1}{4}\hbar^2\right). \tag{7.77}$$

H_{ex} results from the exchange interaction. It is named the Heisenberg exchange interaction, or simply the Heisenberg interaction. Using H_{ex}, the expectation values of H in the singlet and triplet state can now be expressed as

$$\langle\Psi_{\text{singlet}}|H|\Psi_{\text{singlet}}\rangle = 2E_0 + V - \langle\chi_{\text{singlet}}|H_{\text{ex}}|\chi_{\text{singlet}}\rangle, \tag{7.78}$$
$$\langle\Psi_{\text{triplet}}|H|\Psi_{\text{triplet}}\rangle = 2E_0 + V + \langle\chi_{\text{triplet}}|H_{\text{ex}}|\chi_{\text{triplet}}\rangle. \tag{7.79}$$

7.5 ENTANGLEMENT

7.5.1 DENSITY MATRIX

In quantum mechanics, information is embedded in each quantum state, $|\Psi\rangle$, from which the expectation value of any physical observable, O, can be determined from the formula

$$\langle O\rangle = \langle\Psi|O|\Psi\rangle. \tag{7.80}$$

Here $|\Psi\rangle$ is normalized. This equation can be also expressed as

$$\langle O\rangle = \text{Tr}(\rho O), \tag{7.81}$$

where ρ is called the density matrix, defined by

$$\rho = |\Psi\rangle\langle\Psi|. \tag{7.82}$$

Eq. (7.81) implies that the expectation value of O can be also determined from the density matrix, instead of the state or wave function itself.

In this quantum state $|\Psi\rangle$, to measure a physical quantity A is to project this state onto an eigenstate of O, $|n\rangle$. The corresponding probability is given by

$$|\langle n|\Psi\rangle|^2 = \text{Tr}(\rho|n\rangle\langle n|) = \langle n|\rho|n\rangle. \tag{7.83}$$

Like other operators, ρ is a matrix in a particular basis representation. For example, in a spin-1/2 system, there are two linear independent basis states, $|\uparrow\rangle$ and $|\downarrow\rangle$, which are the two eigenstates of σ_z. In this system, $|\Psi\rangle$ is generally a superposition of these two basis states

$$|\Psi\rangle = c_\uparrow|\uparrow\rangle + c_\downarrow|\downarrow\rangle = \sum_{\sigma=\uparrow,\downarrow} c_\sigma|\sigma\rangle. \tag{7.84}$$

This wave function is normalized when $|c_\uparrow|^2 + |c_\downarrow|^2 = 1$. The corresponding density matrix is given by

$$\rho = \sum_{\sigma\sigma'} c^*_{\sigma'}c_\sigma|\sigma\rangle\langle\sigma'|, \tag{7.85}$$

which may be cast into the form

$$\rho = \begin{pmatrix} c_\uparrow & c_\downarrow \end{pmatrix} \begin{pmatrix} c_\uparrow^* \\ c_\downarrow^* \end{pmatrix} = \begin{pmatrix} c_\uparrow^* c_\uparrow, & c_\uparrow^* c_\downarrow \\ c_\downarrow^* c_\uparrow, & c_\downarrow^* c_\downarrow \end{pmatrix} \tag{7.86}$$

in the basis space of σ_z. The diagonal elements of ρ are the square moduli of the coefficients of the two basis states, and their sum equals to 1, hence

$$\mathrm{Tr}\rho = 1 \tag{7.87}$$

as a result of the normalization of wave function. The off-diagonal terms of ρ reflect the interference between the up and down spins. These terms are absent in the corresponding classical system. If we ignore these off-diagonal terms, ρ then becomes

$$\rho_m = |c_\uparrow|^2 |\uparrow\rangle\langle\uparrow| + |c_\downarrow|^2 |\downarrow\rangle\langle\downarrow| = \begin{pmatrix} |c_\uparrow|^2, & 0 \\ 0, & |c_\downarrow|^2 \end{pmatrix}. \tag{7.88}$$

This is like in a classical statistical system, and $|c_\uparrow|^2$ and $|c_\downarrow|^2$ are the probability in the corresponding classical states.

A difference between ρ and ρ_m is given by the formula

$$\mathrm{Tr}\rho^2 \;\;=\;\; \mathrm{Tr}\rho = 1, \tag{7.89}$$
$$\mathrm{Tr}\rho_m^2 \;\;<\;\; \mathrm{Tr}\rho_m = 1. \tag{7.90}$$

This is in fact the defining formula for the pure or mixed states. We call a system a pure state if its density satisfies Eq. (7.89) or a mixed state if its density matrix satisfies Eq. (7.90). Here we use the subscript m in ρ_m to emphasize that it is a density matrix for a mixed state. In either case, the expectation value of physical observable can be found using the formula defined in Eq. (7.81).

The density matrix enlarges the concept of state from a vector in a Hilbert space to an operator from which all properties of a quantum system can be extracted. It is particularly useful in extracting physical properties of a subsystem.

7.5.2 ENTANGLED STATE

In classical mechanics, two or more systems that do not interact, i.e. that do not exert any kind of force on each other, are completely separated, and experiments performed on one of them cannot in any way influence experiments performed on the other. Quantum mechanics, on the other hand, admits that there can be a form of interdependence even in the absence of physical interaction. In other words, the wave function of the whole system cannot be factorized as a product of the wave functions for all subsystems. This type of correlation is called entanglement.

For a system which is composed of two subsystems, if $\{|a_n\rangle, n = 1,\ldots,N\}$ and $\{|b_m\rangle, m = 1,\ldots,M\}$ are the two basis sets of these two subsystems, then the wave function of the system can be generally expressed as

$$|\Psi\rangle = \sum_{nm} \Psi_{nm} |a_n\rangle |b_m\rangle. \tag{7.91}$$

In Eq. (7.91), the wave function Ψ_{nm} is a $N \times M$ matrix in this representation. It can be diagonalized with two unitary matrices by taking a singular value decomposition

$$\Psi = U^T \lambda V, \tag{7.92}$$

where U is a $N \times N$ unitary matrix and V is a $M \times M$ unitary matrix. λ is a semi-definite $N \times M$ diagonal matrix, and

$$\sum_l \lambda_l^2 = \langle \Psi | \Psi \rangle = \text{Tr} \Psi \Psi^\dagger = 1. \tag{7.93}$$

We order λ_l in a descending order, such that $\lambda_i \leq \lambda_j$ if $i > j$. Eq. (7.92) can be also expressed as

$$\Psi_{mn} = \sum_{l=1}^{\min(N,M)} U_{ln} \lambda_l V_{lm}. \tag{7.94}$$

Substituting it into Eq. (7.91), we obtain

$$\boxed{|\Psi\rangle = \sum_l \lambda_l |\tilde{a}_l\rangle |\tilde{b}_l\rangle,} \tag{7.95}$$

where

$$|\tilde{a}_l\rangle = \sum_n U_{ln} |a_n\rangle, \tag{7.96}$$

$$|\tilde{b}_l\rangle = \sum_m V_{lm} |b_m\rangle \tag{7.97}$$

are two new basis sets of the two subsystems.

Eq. (7.95) is called Schmidt decomposition, and the number of nonzero coefficients λ_l is called the Schmidt number. λ_l are the singular values of the wave function Ψ. Their squares are the eigenvalues of the Hermitian matrices $\Psi \Psi^\dagger$ and $\Psi^\dagger \Psi$, whose eigenvectors are the row vectors of U and V, respectively.

If the Schmidt number is 1, namely only the first singular value is finite, then $\lambda_1 = 1$ and

$$|\Psi\rangle = |\tilde{a}_1\rangle |\tilde{b}_1\rangle = \left(\sum_n U_{1n} |a_1\rangle \right) \left(\sum_m V_{1m} |b_1\rangle \right), \tag{7.98}$$

which is simply a product state. We call this a disentangled state or a pure state. Hence $|\Psi\rangle$ is disentangled if its Schmidt number is 1. On the other hand, if the Schmidt number is more than 1, then this state is entangled and cannot be treated as a simple product of separate systems. In other words, an entangled state is a state of a composite system whose subsystems are not probabilistically independent.

7.5.3 ENTANGLEMENT ENTROPY

In order to quantify the degree of entanglement of an entangled state, we introduce the density matrix for a subsystem, called reduced density matrix. For a bipartite system, the reduced density matrix of the first subsystem is defined by tracing out all the degrees of freedom of the second subsystem

$$\rho^{(1)}_{n_1,n_2} = \sum_m \Psi^*_{n_2,m} \Psi_{n_1,m}. \tag{7.99}$$

It can be also expressed as

$$\rho^{(1)} = \Psi\Psi^\dagger. \tag{7.100}$$

Substituting Eq. (7.94) into Eq. (7.99), we obtain

$$\rho^{(1)}_{n_1,n_2} = \sum_{ll'm} U^*_{l,n_2} \lambda_l V^*_{l,m} U_{l',n_1} \lambda_{l'} V_{l',m} = \sum_l U^*_{l,n_2} \lambda_l^2 U_{l,n_1}, \tag{7.101}$$

which can be also expressed as

$$\rho^{(1)} = \sum_l \lambda_l^2 |\tilde{a}_l\rangle\langle\tilde{a}_l| = \mathrm{Tr}_2\rho. \tag{7.102}$$

The entanglement spectra of $\rho^{(1)}$ are given by λ_l^2.

Similarly, we can define the reduced density matrix for the second subsystem

$$\rho^{(2)}_{m_1,m_2} = \sum_n \Psi^*_{n,m_2} \Psi_{n,m_1}, \tag{7.103}$$

or in the matrix form

$$\rho^{(2)} = \Psi^\dagger\Psi = \sum_l \lambda_l^2 |\tilde{b}_l\rangle\langle\tilde{b}_l|. \tag{7.104}$$

$\rho^{(2)}$ has the same entanglement spectrum as $\rho^{(1)}$.

A quantity that is used to measure the degree of entanglement of quantum state $|\Psi\rangle$ is the von Neumann entropy defined using the reduced density matrix

$$\boxed{S = -\mathrm{Tr}\rho^{(1)} \ln\rho^{(1)} = -\mathrm{Tr}\rho^{(2)} \ln\rho^{(2)}.} \tag{7.105}$$

Using the eigen-spectra of the reduced density matrix, it can be written as

$$S = -\sum_l \lambda_l^2 \ln\lambda_l^2. \tag{7.106}$$

For a disentangled state, $\lambda_l = \delta_{l,1}$, $S = 0$. The maximal entanglement entropy emerges when all λ_l have the same value. In this case, the entanglement entropy

$$S_{max} = \ln|\min(N,M)|, \tag{7.107}$$

and the corresponding state is called a maximal entangled state.

An example of a maximally entangled state is the spin singlet state of two $S = 1/2$ spins whose wave function is defined by

$$|\Psi\rangle = \frac{1}{\sqrt{2}} (|\uparrow\downarrow\rangle - |\downarrow\uparrow\rangle). \tag{7.108}$$

The corresponding entanglement entropy equals $\ln 2$.

7.5.4 BELL BASES

Besides the singlet state, defined in Eq. (7.108), two $S = 1/2$ spins can form other three maximally entangled states. These four states are referred to as the Bell bases. They are defined by

$$|\text{Bell 1}\rangle = \frac{1}{\sqrt{2}}(|\uparrow\uparrow\rangle + |\downarrow\downarrow\rangle), \tag{7.109}$$

$$|\text{Bell 2}\rangle = \frac{1}{\sqrt{2}}(|\uparrow\uparrow\rangle - |\downarrow\downarrow\rangle), \tag{7.110}$$

$$|\text{Bell 3}\rangle = \frac{1}{\sqrt{2}}(|\uparrow\downarrow\rangle + |\downarrow\uparrow\rangle), \tag{7.111}$$

$$|\text{Bell 4}\rangle = \frac{1}{\sqrt{2}}(|\uparrow\downarrow\rangle - |\downarrow\uparrow\rangle). \tag{7.112}$$

They form an orthonormal and complete basis set. Any quantum state of two S=1/2 spins can be expanded using these basis states.

7.5.5 EPR PARADOX

The EPR paradox[1] is a thought experiment proposed by A. Einstein, B. Podolsky, and N. Rosen (EPR), that was designed to prove that quantum mechanics is an incomplete theory. This paradox is one of the typical examples of quantum entanglement. A thorough understanding to it would have important implications for the interpretation of quantum mechanics.

In 1951, D. Bohm[2] modified the EPR thought experiment with a simpler example so that the paradox could be more easily understood. The experiment involves an unstable particle of spin 0 which decays into two particles, heading in opposite directions. These two particles are in a spin singlet state, whose wave function is defined by Eq. (7.108), due to the angular momentum conservation.

At the destinations, the two spins are measured independently by two observers. As the two spins are maximally entangled, the measured spin values obtained by the two observers will always be opposite regardless of the relative timing of measurement — if one of the particle is in the up-spin state, the other must be in the down-spin state, and vice versa. It seems that the two measurements are always correlated no matter how far apart the two particles are separated and how short of the time interval between the two measurements. This implies that the first measurement could instantaneously affect the second one. It further implies that information could be transmitted faster than light, violating the principle of special relativity.

If there are many decayed pairs of particles, one would see an odd mixture of probabilistic but inherently correlated results: the sequence of spins measured by

---

[1] A. Einstein, B Podolsky, N Rosen, Can Quantum-Mechanical Description of Physical Reality be Considered Complete? Phys. Rev. **47**, 777 (1935)

[2] D. Bohm, Quantum Theory, Prentice-Hall, Englewood Cliffs, 1951.

each observer vary randomly, but at each time the two spins obtained by the two observers are exactly opposite. The sequence of entangled pairs behaves like a pair of magic coins that one lands head and the other lands tail when tossed together, but which one lands head is completely random. The EPR paradox creates the impression that there is a communication between the two particles at speeds greater than the speed of light. This is in fact incorrect because no communication is needed between two entangled particles. This instantaneous correlation between entangled particles is a natural consequence of quantum superposition of two or more particles, i.e. entanglement, and the collapse of entanglement upon quantum measurement.

The EPR paradox suggests that quantum mechanics is a nonlocal theory, which seemly violates the principle of locality that an object is directly influenced only by its immediate surroundings. To heal this nonlocality, EPR suggested that there might exist a hidden variable that governs the measurement result of two entangled particles. Furthermore, they suggested that the hidden variables are internal to each of the particles and do not depend on the state of faraway particles or measuring devices. The reason we see random instead of deterministic results is because there is no way of accessing this hidden variable. This is a basic assumption of the so called local hidden-variable theory.

In 1964, Bell argued that it is possible to design certain experiments to distinguish quantum mechanics from a local hidden-variable theory. Particularly, he showed that if measurements were performed independently on the two separated particles, then the assumption that the outcomes depend upon hidden variables would impose a constraint on the probabilistic distribution of measurement results. This constraint is formulated as an inequality, known as Bell's inequality. It states that no physical theory of local hidden variables can ever reproduce all of the predictions of quantum mechanics. The Bell's inequality has been tested experimentally many times since 1972. It has been found that the hypothesis of local hidden variables is inconsistent with the way that quantum systems behave.

7.6 PROBLEMS

1. Suppose you had three noninteracting and equal mass particles, in a one-dimensional harmonic oscillator potential, with a total energy $E = 9\hbar\omega/2$.

   a. If they are distinguishable, what are the possible occupation number configurations? What is the most probable configuration? If you picked a particle at random and measured its energy, what values might you get, and what is the probability of each one? What is the most probable energy?

   b. If the particles are three identical fermions, what are the answers to the questions in (a)?

   c. If the particles are three identical bosons, what are the answers to the questions in (a)?

2. Two Bosons reside in the external potential $V(x_1, x_2) = \frac{1}{2}\mu\omega^2(x_1^2 + x_2^2)$, and interact via $U(x_1, x_2) = \lambda(x_1 - x_2)^2$. Compute the ground state energy of the system. What if the two particles are Fermions?

3. Find the entanglement entropy for two spins on the following states

   a. $(|\uparrow\uparrow\rangle + |\downarrow\downarrow\rangle)/\sqrt{2}$

   b. $(|\uparrow\rangle + |\downarrow\rangle)\sqrt{2} \otimes (|\uparrow\rangle + |\downarrow\rangle)/\sqrt{2}$

4. Consider two identical particles (ignoring spin) in a three-dimensional isotropic harmonic oscillator potential,

   a. if the particles are bosons, find the ground state energy and the corresponding eigenfunction.

   b. if the particles are fermions, find the ground state energy and the corresponding eigenfunction.

5. Two identical particles of mass μ are trapped in an infinite square well of width L. If the total energy of the two particles is $5\pi^2\hbar^2/2\mu L^2$, write down the two-particle wave function in each of the following cases:

   a. the particles are spinless;

   b. the particles are electrons;

   c. In (b), if the Coulomb repulsion between two electrons is taken into account, which spin state has the lower energy?

6. Show that the ground state wave function of N Fermions in the lowest Landau level can have the form

$$\Psi(z_1, z_2, \ldots, z_N) = \prod_{i<j}^{N}(z_i - z_j)\exp\left(-\frac{1}{4\ell^2}\sum_{i=1}^{N}|z_i|^2\right),$$

where $z_i = x_i + iy_i$ is the coordinate of the Fermions and $\ell = \sqrt{\frac{\hbar c}{eB}}$ is the magnetic length.

8 Symmetry and Conservation Law

8.1 SPATIAL TRANSLATION INVARIANCE AND MOMENTUM CONSERVATION

If a system is invariant under certain transformation, we say that the system is symmetric under that transformation. In other words, a symmetry is an equivalence of different physical states. Corresponding to each symmetry, there is a conservation law with a characteristic conserving quantity which is invariant under that transformation. One of the aims of physical research is to search for symmetries and the corresponding observables that are invariant under certain classes of transformation. In this context, it becomes particularly interesting to look for the conditions under which a certain symmetry is broken.

In Section 2.1, we consider the parity (symmetry) of wave function under the spatial reflection. Now let us consider the motion of a single particle under an infinitesimal translation in one dimension.

8.1.1 TRANSLATION OPERATOR

We use $T(\varepsilon)$ to denote an infinitesimal translation operator with ε the amount of shift in space. Under this transformation, a quantum state is transformed to

$$|\Psi_\varepsilon\rangle = T(\varepsilon)|\Psi\rangle. \tag{8.1}$$

Correspondingly, the position of the particle is changed to

$$\langle\Psi_\varepsilon|x|\Psi_\varepsilon\rangle = \langle\Psi|x|\Psi\rangle + \varepsilon. \tag{8.2}$$

Both $|\Psi_\varepsilon\rangle$ and $|\Psi\rangle$ are normalized.

The above equation can be also represented as

$$\langle\Psi|T^\dagger(\varepsilon)xT(\varepsilon)|\Psi\rangle = \langle\Psi|x|\Psi\rangle + \varepsilon. \tag{8.3}$$

Since this equation is valid for arbitrary states, we obtain the transformation formula for the position operator

$$\boxed{T^\dagger(\varepsilon)xT(\varepsilon) = x + \varepsilon.} \tag{8.4}$$

In real space, $T(\varepsilon)$ is to transform state $|x\rangle$ to $|x+\varepsilon\rangle$

$$\boxed{T(\varepsilon)|x\rangle = |x+\varepsilon\rangle.} \tag{8.5}$$

DOI: 10.1201/9781003174882-8

The wave function is transformed as

$$
\begin{aligned}
\langle x|T(\varepsilon)|\Psi\rangle &= \int dx' \langle x|T(\varepsilon)|x'\rangle \langle x'|\Psi\rangle \\
&= \int dx' \langle x|x'+\varepsilon\rangle \Psi(x') \\
&= \Psi(x-\varepsilon).
\end{aligned}
\tag{8.6}
$$

In other words,

$$
T(\varepsilon)\Psi(x) = \Psi(x-\varepsilon).
\tag{8.7}
$$

Using Eq. (8.5), we find that

$$
\langle x'|T^{\dagger}(\varepsilon)T(\varepsilon)|x\rangle = \langle x'+\varepsilon|x+\varepsilon\rangle = \delta(x'-x).
\tag{8.8}
$$

It indicates that $T(\varepsilon)$ is a unitary transformation

$$
T^{\dagger}(\varepsilon)T(\varepsilon) = I.
\tag{8.9}
$$

8.1.2 GENERATOR OF TRANSLATIONS

As $\varepsilon = 0$ corresponds to no translation, we may expand $T(\varepsilon)$ to order ε as

$$
\boxed{T(\varepsilon) = I - \frac{i\varepsilon}{\hbar}P \approx e^{-i\varepsilon P/\hbar}.}
\tag{8.10}
$$

I is the identity operator. Operator P is called the generator of translations. Since $T(\varepsilon)$ is unitary, we have

$$
T^{\dagger}(\varepsilon)T(\varepsilon) = \left(I + \frac{i\varepsilon}{\hbar}P^{\dagger}\right)\left(I - \frac{i\varepsilon}{\hbar}P\right) = I + \frac{i\varepsilon}{\hbar}(P^{\dagger}-P) + o(\varepsilon^2) = I.
\tag{8.11}
$$

Thus $P = P^{\dagger}$ is a Hermitian operator.

Furthermore, to act $T(\varepsilon)$ on a wave function in real space, i.e. $\Psi(x) = \langle x|\Psi\rangle$, we find that

$$
\langle x|T(\varepsilon)|\Psi\rangle = \langle x|1 - \frac{i\varepsilon}{\hbar}P|\Psi\rangle = \Psi(x) - \frac{i\varepsilon}{\hbar}\langle x|P|\Psi\rangle.
\tag{8.12}
$$

Equating it with the Taylor expansion of the wave function (8.6),

$$
\langle x|T(\varepsilon)|\Psi\rangle = \Psi(x-\varepsilon) \approx \Psi(x) - \varepsilon\frac{\partial}{\partial x}\Psi(x),
\tag{8.13}
$$

we find that

$$
\langle x|P|\Psi\rangle = -i\hbar\frac{\partial}{\partial x}\Psi(x).
\tag{8.14}
$$

It shows that the generator of translation P is nothing but the momentum operator

$$
\boxed{P = -i\hbar\frac{\partial}{\partial x}.}
\tag{8.15}
$$

This expression can be also regarded as a defining equation of the momentum operator.

8.1.3 MOMENTUM CONSERVATION

If the Hamiltonian is invariant under the spatial translation, namely

$$T^\dagger(\varepsilon)HT(\varepsilon) = H, \tag{8.16}$$

we say that the system is translation invariant.

Substituting Eq. (8.10) into Eq. (8.16), we obtain

$$0 = T^\dagger(\varepsilon)HT(\varepsilon) - H = \frac{i\varepsilon}{\hbar}[P,H] + O(\varepsilon^2), \tag{8.17}$$

so that

$$\boxed{[P,H] = 0.} \tag{8.18}$$

It implies that $\partial P/\partial t = 0$ and the momentum is conserved. Thus the translation invariance corresponds to the conservation of momentum.

8.1.4 FINITE TRANSLATION

The operator $T(x)$ corresponding to a finite translation x can be found by dividing x into N parts of size x/N. In the limit $N \to \infty$, x/N becomes infinitesimal and $T(x/N)$ is given by Eq. (8.10). Since a translation by x equals N translation by x/N,

$$\boxed{T(x) = \lim_{N\to\infty} T^N\left(\frac{x}{N}\right) = \exp\left(-\frac{ix}{\hbar}P\right) = \exp(-x\partial_x)} \tag{8.19}$$

by virtue of the formula

$$e^{-x} = \lim_{N\to\infty}\left(1 - \frac{x}{N}\right)^N. \tag{8.20}$$

A generalization of Eq. (8.7) at finite transition is

$$T(a)\Psi(x) = \Psi(x-a). \tag{8.21}$$

It is simple to verify this equation by taking Taylor expansion

$$T(a)\Psi(x) = e^{-a\partial_x}\Psi(x) = \sum_n \frac{(-a)^n}{n!}\partial_x^n\Psi(x) = \Psi(x-a). \tag{8.22}$$

For an operator $O(x)$, it changes under an infinitesimal translation according to the formula

$$
\begin{aligned}
T(\varepsilon)O(x)T^\dagger(\varepsilon) &\approx (1-\varepsilon\partial_x)O(x)(1+\varepsilon\partial_x) \\
&= O(x) - \varepsilon\partial_x O(x) + o(\varepsilon^2) \\
&= O(x-\varepsilon) + o(\varepsilon^2).
\end{aligned} \tag{8.23}
$$

In general,

$$T(a)O(x)T^\dagger(a) = O(x-a). \tag{8.24}$$

8.1.5 STONE THEOREM

From the above discussion, we see that the translational operator can be cast into the form

$$T(x) = e^{-ixP/\hbar}. \tag{8.25}$$

This exponential expression of unitary transformation operator actually holds rather generally. A more general statement was proven by Stone[1], which is referred to as a theorem named after him.

Stone Theorem: Given a family $U(\alpha)$ of unitary operators, where α is a real parameter, satisfying the property of Abelian group

$$U(\alpha)U(\beta) = U(\alpha + \beta), \tag{8.26}$$

then it is possible to write

$$U(\alpha) = \exp(i\alpha G), \tag{8.27}$$

where G is a Hermitian operator.

This theorem establishes a one-to-one correspondence between a self-adjoint operator on a Hilbert space and a one-parameter unitary and continuous transformation. It may be proven following the step for deriving the expression of finite translation operator $T(x)$ in Section 8.1.

In Eq. (8.27), G is the generator of the corresponding unitary transformation $U(a)$. The Stone theorem ensures the existence of a Hermitian infinitesimal generator for an Abelian group of unitary transformation. The spatial translation operator is a specific example of this theorem and the momentum operator is the generator of spatial translation.

8.2 GALILEAN INVARIANCE

In classical mechanics, the equation of motion is exactly the same in two inertial coordinate systems, I and I', moving relative to each other with uniform velocity $\mathbf{v}$ (Fig. 8.1). This is called Galilean invariance. The coordinate and momentum in these two coordinate systems are transformed according to the equations

$$\begin{aligned} \mathbf{r}' &= \mathbf{r} - \mathbf{v}t, &(8.28) \\ \mathbf{p}' &= \mathbf{p} - \mu\mathbf{v}. &(8.29) \end{aligned}$$

Neither coordinate system is preferred.

This Galilean principle of relativity must be true in quantum mechanics as well. Indeed, the Schrödinger equation is invariant under the Galilean transformaion. To

---

[1] M. H. Stone, Linear transformations in Hilbert space, III: operational methods and group theory, Proc. Nat. Acad. Sci. U.S.A. **16**, 172 (1930).

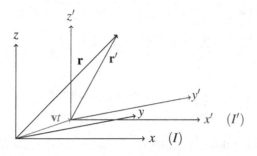

Figure 8.1 Reference I' moves with respect to reference I in a uniform velocity **v**.

prove this, let us assume $\Psi(\mathbf{r},t)$ and $\Psi'(\mathbf{r}',t)$ to be the wave functions in the two reference systems, and $\Psi(\mathbf{r},t)$ is a solution of the Schrödinger equation.

$$i\hbar\frac{\partial}{\partial t}\Psi(\mathbf{r},t) = -\frac{\hbar^2}{2\mu}\nabla^2\Psi(\mathbf{r},t) + V(\mathbf{r})\Psi(\mathbf{r},t). \tag{8.30}$$

If $\Psi'(\mathbf{r},t)$ describes the same quantum state, it should be equal to $\Psi(\mathbf{r}+\mathbf{v}t,t)$ up to a phase factor

$$\Psi'(\mathbf{r},t) = e^{i\theta(\mathbf{r},t)/\hbar}\Psi(\mathbf{r}+\mathbf{v}t,t) = e^{i\theta(\mathbf{r},t)/\hbar}e^{i\mathbf{p}\cdot\mathbf{v}t/\hbar}\Psi(\mathbf{r},t). \tag{8.31}$$

$\theta(\mathbf{r},t)$ is determined by requiring that $\Psi'(\mathbf{r},t)$ is a solution of the Schrödinger equation in the I'-coordinate system

$$i\hbar\frac{\partial}{\partial t}\Psi'(\mathbf{r},t) = -\frac{\hbar^2}{2\mu}\nabla^2\Psi'(\mathbf{r},t) + V(\mathbf{r}+\mathbf{v}t)\Psi'(\mathbf{r},t). \tag{8.32}$$

To determine $\theta(\mathbf{r},t)$, we first evaluate the time and spatial derivatives of $\Psi'(\mathbf{r},t)$. It is simple to show that

$$i\hbar\partial_t\Psi'(\mathbf{r},t) = e^{i\theta(\mathbf{r},t)/\hbar}\left[i\hbar\partial_t - (\partial_t\theta)\right]e^{i\mathbf{p}\cdot\mathbf{v}t/\hbar}\Psi(\mathbf{r},t) \tag{8.33}$$

As **p** commutes with $\exp(i\mathbf{p}\cdot\mathbf{v}t/\hbar)$, the spatial derivative of $\Psi'(\mathbf{r},t)$ is

$$-i\hbar\nabla\Psi'(\mathbf{r},t) = e^{i\theta(\mathbf{r},t)/\hbar}\left[-i\hbar\nabla + (\nabla\theta)\right]e^{i\mathbf{p}\cdot\mathbf{v}t/\hbar}\Psi(\mathbf{r},t), \tag{8.34}$$

hence

$$\mathbf{p}\Psi'(\mathbf{r},t) = e^{i\theta(\mathbf{r},t)/\hbar}\left[\mathbf{p} + (\nabla\theta)\right]e^{i\mathbf{p}\cdot\mathbf{v}t/\hbar}\Psi(\mathbf{r},t), \tag{8.35}$$

and

$$\mathbf{p}^2\Psi'(\mathbf{r},t) = e^{i\theta(\mathbf{r},t)/\hbar}\left[\mathbf{p} + (\nabla\theta)\right]^2 e^{i\mathbf{p}\cdot\mathbf{v}t/\hbar}\Psi(\mathbf{r},t). \tag{8.36}$$

Inserting the above derivatives to Eq. (8.32) yields

$$\left[i\hbar\partial_t - (\partial_t\theta)\right]e^{i\mathbf{p}\cdot\mathbf{v}t/\hbar}\Psi(\mathbf{r},t)$$
$$= \frac{1}{2\mu}\left[\mathbf{p} + (\nabla\theta)\right]^2 e^{i\mathbf{p}\cdot\mathbf{v}t/\hbar}\Psi(\mathbf{r},t) + V(\mathbf{r}+\mathbf{v}t)e^{i\mathbf{p}\cdot\mathbf{v}t/\hbar}\Psi(\mathbf{r},t). \tag{8.37}$$

It can be rewritten as

$$(i\hbar\partial_t - \mathbf{p}\cdot\mathbf{v})\,\Psi(\mathbf{r},t) - \underline{\partial_t\theta}\Psi(\mathbf{r},t) = \frac{1}{2\mu}\,(\mathbf{p}+\underline{\nabla\theta})^2\,\Psi(\mathbf{r},t) + V(\mathbf{r})\Psi(\mathbf{r},t), \quad (8.38)$$

where

$$\underline{\partial_t\theta} = e^{-i\mathbf{p}\cdot\mathbf{v}t/\hbar}\,(\partial_t\theta)\,e^{i\mathbf{p}\cdot\mathbf{v}t/\hbar}, \tag{8.39}$$

$$\underline{\nabla\theta} = e^{-i\mathbf{p}\cdot\mathbf{v}t/\hbar}\,(\nabla\theta)\,e^{i\mathbf{p}\cdot\mathbf{v}t/\hbar}. \tag{8.40}$$

As $\Psi(\mathbf{r},t)$ is a solution of Eq. (8.30), this yields the equation that θ satisfies

$$-\underline{\partial_t\theta} - \frac{1}{2\mu}\,(\underline{\nabla\theta})^2 = \frac{1}{2\mu}\,(\mathbf{p}\cdot\underline{\nabla\theta}+\underline{\nabla\theta}\cdot\mathbf{p}) + \mathbf{p}\cdot\mathbf{v}. \tag{8.41}$$

A special solution is determined by demanding both sides of this equation to be zero. This yields

$$\underline{\nabla\theta} = -\mu\mathbf{v}, \tag{8.42}$$

$$\underline{\partial_t\theta} = -\frac{1}{2\mu}\,(\underline{\nabla\theta})^2 = -\frac{\mu\mathbf{v}^2}{2}, \tag{8.43}$$

and

$$\theta(\mathbf{r},t) = -\mu\mathbf{v}\cdot\mathbf{r} - \frac{\mu\mathbf{v}^2 t}{2}, \tag{8.44}$$

up to a time and coordinate constant. More generally, $\theta(\mathbf{r},t)$ can be expressed as

$$\boxed{\theta(\mathbf{r},t) = -\mu\mathbf{v}\cdot\mathbf{r} - \frac{\mu\mathbf{v}^2 t}{2} + \vartheta(\mathbf{r},t),} \tag{8.45}$$

provided it satisfies Eq. (8.41). For simplicity, we set $\vartheta = 0$ for the discussion below.

Thus the Schrödinger equation is invariant under the following Gelilean transformation of the wave function

$$\boxed{\Psi'(\mathbf{r},t) = e^{-\frac{i}{\hbar}\left(\mu\mathbf{v}\cdot\mathbf{r}+\frac{\mu\mathbf{v}^2 t}{2}\right)}\,e^{i\mathbf{p}\cdot\mathbf{v}t/\hbar}\Psi(\mathbf{r},t).} \tag{8.46}$$

Furthermore, it can be shown that the physical quantities satisfy the following properties:

1. The expectation value of any function of the coordinate is transformed according to Eq. (8.28)

$$\begin{aligned}
\langle\Psi'|f(\mathbf{r})|\Psi'\rangle &= \int d\mathbf{r}\,\Psi'^*(\mathbf{r},t)f(\mathbf{r})\Psi'(\mathbf{r},t) \\
&= \int d\mathbf{r}\,\Psi^*(\mathbf{r},t)e^{-i\mathbf{p}\cdot\mathbf{v}t/\hbar}f(\mathbf{r})e^{i\mathbf{p}\cdot\mathbf{v}t/\hbar}\Psi(\mathbf{r},t) \\
&= \int d\mathbf{r}\,\Psi^*(\mathbf{r},t)f(\mathbf{r}-\mathbf{v}t)\Psi(\mathbf{r},t) \\
&= \langle\Psi|f(\mathbf{r}-\mathbf{v}t)|\Psi\rangle. \tag{8.47}
\end{aligned}$$

2. The expectation value of any function of the momentum is transformed according to Eq. (8.29)

$$
\begin{aligned}
\langle \Psi'|g(\mathbf{p})|\Psi'\rangle &= \int d\mathbf{r}\Psi'^*(\mathbf{r},t)g(\mathbf{p})\Psi'(\mathbf{r},t) \\
&= \int d\mathbf{r}\Psi^*(\mathbf{r},t)e^{-i\mathbf{p}\cdot\mathbf{v}t/\hbar}e^{-i\mu\mathbf{v}\cdot\mathbf{r}/\hbar}g(\mathbf{p})e^{i\mu\mathbf{v}\cdot\mathbf{r}/\hbar}e^{i\mathbf{p}\cdot\mathbf{v}t/\hbar}\Psi(\mathbf{r},t) \\
&= \int d\mathbf{r}\Psi^*(\mathbf{r},t)g(\mathbf{p}-\mu\mathbf{v})\Psi(\mathbf{r},t) \\
&= \langle \Psi|g(\mathbf{p}-\mu\mathbf{v})|\Psi\rangle.
\end{aligned}
\tag{8.48}
$$

Here $\exp(i\mu\mathbf{v}\cdot\mathbf{r}/\hbar)$ is a translation operator in momentum space

$$
e^{i\mu\mathbf{v}\cdot\mathbf{r}/\hbar}g(\mathbf{p})e^{-i\mu\mathbf{v}\cdot\mathbf{r}/\hbar} = e^{-i\mu\mathbf{v}\cdot\nabla_{\mathbf{p}}}g(\mathbf{p})e^{i\mu\mathbf{v}\cdot\nabla_{\mathbf{p}}} = g(\mathbf{p}-\mu\mathbf{v}).
\tag{8.49}
$$

8.3 NOETHER THEOREM

From the discussion given in Section 8.1, we know that the translational symmetry implies the conservation of momentum. In general, if a system has a symmetry under certain continuous transformation so that its Hamiltonian is invariant under the transformation, then there is a corresponding physical quantity which is conserved in time. This conserved quantity is nothing but the infinitesimal generator of this symmetry, which commutes with the Hamiltonian and is a constant of motion. This statement is a generalization of the one first proven by Emmy Noether.

> Noether Theorem: Every continuous symmetry has a corresponding constant of motion given by the observable that represents the infinitesimal generator of the corresponding transformation.

To prove this theorem, let us assume $U(\varepsilon)$ to be an infinitesimal transformation operator and the Hamiltonian H is invariant under this transformation

$$
U^\dagger(\varepsilon)HU(\varepsilon) = H.
\tag{8.50}
$$

Using the infinitesimal generator of this transformation G, we can write $U(\varepsilon)$ as

$$
U(\varepsilon) = 1 - \frac{i\varepsilon}{\hbar}G.
\tag{8.51}
$$

Since $U(\alpha)$ is a unitary transformation, G must be hermitian.

Substituting this expression into Eq. (8.50) and keeping the leading order terms in ε, it is simple to show that

$$
\boxed{[G,H] = 0.}
\tag{8.52}
$$

From the Heisenberg equation of G, we know that G is a constant of motion.

8.4 ROTATION AND ANGULAR MOMENTUM CONSERVATION

As an application of the Noether theorem, we consider how the physical states are changed under rotation.

8.4.1 ROTATION IN TWO DIMENSIONS

Let us take a rotation by an angle ϕ about the z axis (counterclockwise in the $x - y$ plane), which changes the coordinate (x, y) to

$$x \;\rightarrow\; x\cos\phi - y\sin\phi, \tag{8.53}$$

$$y \;\rightarrow\; x\sin\phi + y\cos\phi. \tag{8.54}$$

The corresponding rotation operator R can therefore be defined as

$$R[\phi\hat{z}]\,|x,y\rangle = |x\cos\phi - y\sin\phi, x\sin\phi + y\cos\phi\rangle. \tag{8.55}$$

For an infinitesimal rotation, $\phi = \varepsilon$, the rotation operator R could be written as

$$R[\varepsilon\hat{z}] = I - \frac{i\varepsilon}{\hbar}L_z, \tag{8.56}$$

where L_z is the generator of infinitesimal rotation and a hermitian operator.

To determine L_z, let us apply $R[\varepsilon\hat{z}]$ to an arbitrary physical state

$$
\begin{aligned}
\langle x,y|R[\varepsilon\hat{z}]|\Psi\rangle &= \int dx'dy'\,\langle x,y|R[\varepsilon\hat{z}]|x',y'\rangle\langle x',y'|\Psi\rangle \\
&= \int dx'dy'\,\langle x,y|x'\cos\varepsilon - y'\sin\varepsilon, x'\sin\varepsilon + y'\cos\varepsilon\rangle\Psi(x',y') \\
&= \Psi(x\cos\varepsilon + y\sin\varepsilon, -x\sin\varepsilon + y\cos\varepsilon).
\end{aligned}
\tag{8.57}
$$

Expanding both sides to order ε, we obtain

$$\Psi(x,y) - \frac{i\varepsilon}{\hbar}\langle x,y|L_z|\Psi\rangle \approx \Psi(x,y) - \varepsilon\left(x\frac{\partial}{\partial y} - y\frac{\partial}{\partial x}\right)\Psi(x,y). \tag{8.58}$$

It leads to the equality

$$\langle x,y|L_z|\Psi\rangle = -i\hbar\left(x\frac{\partial}{\partial y} - y\frac{\partial}{\partial x}\right)\Psi(x,y), \tag{8.59}$$

hence

$$\boxed{L_z = -i\hbar\left(x\frac{\partial}{\partial y} - y\frac{\partial}{\partial x}\right).} \tag{8.60}$$

The generator of two-dimensional rotational transformation is the third component of the orbital angular momentum, which is conserved if the Hamiltonian is invariant under the rotation. This provides a definition of L_z which justifies the expression $L_z = (\mathbf{r} \times \mathbf{p})_z$ used before.

8.4.2   ROTATION IN THREE DIMENSIONS

In three dimensions, a rotation can be done along any of the three principal axes. Correspondingly, there are three generators, along the x, y, and z-axis, respectively. We define these generators as L_x, L_y, L_z, respectively. The corresponding infinitesimal rotations are

$$R[\varepsilon \hat{x}] = I - \frac{i\varepsilon}{\hbar}L_x, \tag{8.61}$$

$$R[\varepsilon \hat{y}] = I - \frac{i\varepsilon}{\hbar}L_y, \tag{8.62}$$

$$R[\varepsilon \hat{z}] = I - \frac{i\varepsilon}{\hbar}L_z. \tag{8.63}$$

From the discussion on the two-dimensional rotational transformation, we know that L_z is just the third component of the angular momentum. Here, it is natural to expect that L_x, L_y, and L_z are the three operators of the orbital angular momentum, satisfying the commutation relations of angular momenta.

Let us take four infinitesimal rotations, in the order $R[\varepsilon_x \hat{x}]$, $R[\varepsilon_y \hat{y}]$, $R[-\varepsilon_x \hat{x}]$, and lastly $R[-\varepsilon_y \hat{y}]$. Since

$$R[\varepsilon_x \hat{x}] \mid x, \quad y, \quad z \, \rangle \approx \mid x, \quad y - \varepsilon_x z \quad z + \varepsilon_x y \, \rangle, \tag{8.64}$$
$$R[\varepsilon_y \hat{y}] \mid x, \quad y, \quad z \, \rangle \approx \mid x + \varepsilon_y z, \quad y, \quad z - \varepsilon_y x \, \rangle, \tag{8.65}$$

we have

$$
\begin{aligned}
&R[-\varepsilon_y \hat{y}] R[-\varepsilon_x \hat{x}] R[\varepsilon_y \hat{y}] R[\varepsilon_x \hat{x}] \mid x, \quad y, \quad z \, \rangle \\
={} & R[-\varepsilon_y \hat{y}] R[-\varepsilon_x \hat{x}] R[\varepsilon_y \hat{y}] \mid x, \quad y - \varepsilon_x z, \quad z + \varepsilon_x y \, \rangle \\
={} & R[-\varepsilon_y \hat{y}] R[-\varepsilon_x \hat{x}] \mid x + \varepsilon_y z + \varepsilon_x \varepsilon_y y, \quad y - \varepsilon_x z, \quad z + \varepsilon_x y - \varepsilon_y x \, \rangle \\
={} & R[-\varepsilon_y \hat{y}] \mid x + \varepsilon_y z + \varepsilon_x \varepsilon_y y, \quad y - \varepsilon_x \varepsilon_y x + o(\varepsilon_x^2), \quad z - \varepsilon_y x + o(\varepsilon_x^2) \, \rangle \\
\approx{} & \mid x + \varepsilon_x \varepsilon_y y + o(\varepsilon_y^2), \quad y - \varepsilon_x \varepsilon_y x + o(\varepsilon_x^2), \quad z + o(\varepsilon_x^2) + o(\varepsilon_y^2) \, \rangle \\
={} & R\left[-\varepsilon_x \varepsilon_y \hat{z} + o(\varepsilon_x^2) + o(\varepsilon_y^2)\right] \mid x, \quad y, \quad z \, \rangle. \tag{8.66}
\end{aligned}
$$

Hence

$$R[-\varepsilon_y \hat{y}] R[-\varepsilon_x \hat{x}] R[\varepsilon_y \hat{y}] R[\varepsilon_x \hat{x}] = R\left[-\varepsilon_x \varepsilon_y \hat{z} + o(\varepsilon_x^2) + o(\varepsilon_y^2)\right]. \tag{8.67}$$

It can be also expressed as

$$\left(I + \frac{i\varepsilon_y}{\hbar}L_y\right)\left(I + \frac{i\varepsilon_x}{\hbar}L_x\right)\left(I - \frac{i\varepsilon_y}{\hbar}L_y\right)\left(I - \frac{i\varepsilon_x}{\hbar}L_x\right) \tag{8.68}$$

$$= I + \frac{i\varepsilon_x \varepsilon_y}{\hbar}L_z + o(\varepsilon_x^2) + o(\varepsilon_y^2). \tag{8.69}$$

If we expand the above expression to the second order in ε

$$\varepsilon_x \varepsilon_y \left(L_x L_y - L_y L_x\right) + \varepsilon_x^2 L_x^2 + \varepsilon_y^2 L_y^2 \approx i\hbar \varepsilon_x \varepsilon_y L_z + o(\varepsilon_x^2) + o(\varepsilon_y^2). \tag{8.70}$$

Matching the coefficient of $\varepsilon_x \varepsilon_y$, we find that

$$[L_x, L_y] = i\hbar L_z, \tag{8.71}$$

which is just the defining equation of angular momenta.

Similarly, it can be shown that

$$[L_y, L_z] = i\hbar L_x, \tag{8.72}$$
$$[L_z, L_x] = i\hbar L_y. \tag{8.73}$$

Thus the three orbital angular momentum operators are just the three generators of three-dimensional rotations.

8.4.3   ANGULAR MOMENTUM CONSERVATION

The infinitesimal rotation operator can be expressed as

$$R(\varepsilon \hat{n}) = I - \frac{i\varepsilon}{\hbar} \hat{n} \cdot \mathbf{L} = \exp\left(-\frac{i\varepsilon}{\hbar} \hat{n} \cdot \mathbf{L}\right). \tag{8.74}$$

Using the Stone theorem, the rotation operator for an arbitrary angle along $\hat{n}$ is given by

$$R(\phi \hat{n}) = \exp\left(-\frac{i\phi}{\hbar} \hat{n} \cdot \mathbf{L}\right). \tag{8.75}$$

If the Hamiltonian is invariant under arbitrary rotation

$$R^\dagger(\phi \hat{n}) H R(\phi \hat{n}) = H, \tag{8.76}$$

it follows, upon considering infinitesimal rotations around the three axes, that

$$\boxed{[\mathbf{L}, H] = 0.} \tag{8.77}$$

Thus the angular momentum is conserved.

8.5   TIME TRANSLATION INVARIANCE AND ENERGY CONSERVATION

Just as the homogeneity of space ensures that the same experiment performed at two different places gives the same result, homogeneity in time ensures that the same experiment repeated at two different times gives the same result.

Let us prepare at time t_1 a system in state $|\Psi_0\rangle$ and let it evolve for an infinitesimal time ε. The state at time $t_1 + \varepsilon$, to first order in ε, will be

$$|\Psi(t_1 + \varepsilon)\rangle = \left[1 - \frac{i\varepsilon}{\hbar} H(t_1)\right] |\Psi_0\rangle. \tag{8.78}$$

If we repeat the experiment at time t_2, beginning with the same initial state, the state at time $t_2 + \varepsilon$ will be

$$|\Psi(t_2 + \varepsilon)\rangle = \left[1 - \frac{i\varepsilon}{\hbar} H(t_2)\right] |\Psi_0\rangle. \tag{8.79}$$

If the system is time translation invariant, then

$$0 = |\Psi(t_2 + \varepsilon)\rangle - |\Psi(t_1 + \varepsilon)\rangle = -\frac{i\varepsilon}{\hbar} [H(t_2) - H(t_1)] |\Psi_0\rangle. \tag{8.80}$$

Since $|\Psi_0\rangle$ is arbitrary, it follows that

$$H(t_2) = H(t_1), \tag{8.81}$$

hence H is time-independent

$$\boxed{\frac{\partial}{\partial t} H(t) = 0.} \tag{8.82}$$

Thus time translational invariance requires the Hamiltonian to be time independence, which implies that the energy $\langle H \rangle$ is conserved.

8.6 TIME-REVERSAL SYMMETRY

Time reversal is a discrete transformation. It is to convert a quantum state $|\Psi\rangle$ into its time-reversed state. If T is the time-reversal operator and $|\Psi\rangle = |p\rangle$ is a momentum eigenstate, $T|p\rangle$ is equal to $|-p\rangle$ up to a possible phase factor.

Time-reversal symmetry is a symmetry of the system under the transformation of time reversal $T : t \to -t$. This symmetry does not exist in a macroscopically open system since its entropy always increases with time according to the second law of thermodynamics. Hence, it is purely a property of equilibrium system whose coupling with environment is insulated.

Time reversal is an important concept in physics. Quite many interesting physical effects are associated with the time-reversal symmetry. For example, The non-trivial topological order of a topological insulator, which is a material that behaves as an insulator in its interior but a metal on its surface, is protected by the time-reversal symmetry.

The time reversal changes t to $-t$. However, It should be emphasized that T is not to reverse the direction of time flow. Instead, it is to change t to $-t$ and reverse the direction of "motion" at each given time. The time evolution is still along the direction of increasing time after the time-reversal transformation.

The time-reversal transformation is peculiar. It shows two features that are not observed in the rotation or other spatial transformations:

1. If a quantum state evolves from an initial state $|\Psi_i\rangle$ to a final state $|\Psi_f\rangle$ at time t, under the time-reversal translation, these states change to $T|\Psi_i\rangle$

and $T|\Psi_f\rangle$, respectively. In addition to this, the order of the initial and final states are changed. This suggests that

$$\boxed{\langle T\Psi_f|T\Psi_i\rangle = \langle \Psi_f|T^\dagger T|\Psi_i\rangle = \langle \Psi_i|\Psi_f\rangle = \langle \Psi_f|\Psi_i\rangle^*} \qquad (8.83)$$

if the system is time-reversal symmetric. This equation implies that T is not a unitary transformation, and consequently there is no conservation law associated with the time-reversal symmetry.

2. The time evolution of the quantum state is governed by the Schrödinger equation

$$i\hbar\frac{\partial}{\partial t}\Psi(t) = H\Psi(t). \qquad (8.84)$$

Given a solution $\Psi(t)$, it is apparently that $\Psi(-t)$ generally does not satisfies the equation, due to the reverse of time. However, $\Psi^*(-t)$ satisfies this equation if H is time-independent

$$i\hbar\frac{\partial}{\partial t}\Psi^*(-t) = H\Psi^*(-t). \qquad (8.85)$$

It again suggests that the time-reversal transformation is not unitary. Furthermore, it indicates that the time reversal involves the operation of complex conjugation.

8.6.1 TIME-REVERSAL TRANSFORMATION

We now deduce the property of the time-reversal operator T by examining the time evolution of the time-reversed state. As already mentioned, the time-reversal transformation changes the direction of motion at each given time, but it does not change the direction of time evolution. This means that if we start from a time-reversal state $|\tilde{\Psi}_0\rangle = T|\Psi(t=0)\rangle$ at time $t = 0$, this state will evolves into a state at time $t = \tau$ that obeys the Schrödinger equation without changing the direction of time, namely the state

$$|\tilde{\Psi}(\tau)\rangle = e^{-i\tau H/\hbar}|\tilde{\Psi}_0\rangle = e^{-i\tau H/\hbar}T|\Psi(0)\rangle. \qquad (8.86)$$

This state should equal the time-reversed state at time $t = -\tau$, $T|\Psi(-\tau)\rangle$, if the system is time-reversal symmetric

$$|\tilde{\Psi}(\tau)\rangle = T|\Psi(-\tau)\rangle, \qquad (8.87)$$

hence

$$e^{-i\tau H/\hbar}T|\Psi(0)\rangle = Te^{i\tau H/\hbar}|\Psi(0)\rangle. \qquad (8.88)$$

In the limit $\tau \to 0$, it becomes

$$\left(1 - \frac{i\tau H}{\hbar}\right)T|\Psi_0\rangle = T\left(1 + \frac{i\tau H}{\hbar}\right)|\Psi_0\rangle. \qquad (8.89)$$

Equalizing the coefficients of τ on the two sides of the equation yields

$$TiH|\Psi_0\rangle = -iHT|\Psi_0\rangle. \tag{8.90}$$

Here i's on the two sides of the equation should not be cancelled. Otherwise, we would have

$$TH = -HT. \tag{8.91}$$

This equation is physically incorrect because if $|n\rangle$ is an eigenstate of H with an eigenvalue E_n, then $T|n\rangle$ is also an eigenstate of H with the eigenvalue $-E_n$

$$H(T|n\rangle) = -TH|n\rangle = -E_n(T|n\rangle). \tag{8.92}$$

It means that the eigen-spectrum of H must be always symmetric with respect to the zero energy, which is clearly incorrect because the energy spectrum of a free particle (which is clearly time-reversal invariant) is positive definite and there is no state lower than a particle at rest whose energy is zero.

The above argument suggests that T should commute with H

$$\boxed{TH = HT,} \tag{8.93}$$

but anti-commute with i

$$\boxed{Ti = -iT.} \tag{8.94}$$

This is a fascinating equation. It shows that i is not simply a mathematical number, rather it is a "physical variable" that can be "measured" by the time-reversal transformation.

The above equation shows that T is not a unitary operator. In fact, it is an anti-unitary operator.

Definition: A transformation T is said to be antiunitary if

$$\begin{cases} \langle T\phi|T\psi\rangle = \langle\phi|\psi\rangle^*, \\ T(a|\psi\rangle + b|\phi\rangle) = a^*T|\psi\rangle + b^*T|\phi\rangle. \end{cases} \tag{8.95}$$

An antiunitary operator T can be written as a product of a unitary operator U and a complex-conjugate operator K

$$\boxed{T = UK.} \tag{8.96}$$

A proof for this is simple:

$$\langle T\phi|T\psi\rangle = \langle K\phi|U^\dagger U|K\psi\rangle = \langle\phi^*|\psi^*\rangle = \langle\phi|\psi\rangle^*, \tag{8.97}$$

and

$$\begin{aligned} T(a|\psi\rangle + b|\phi\rangle) &= UK(a|\psi\rangle + b|\phi\rangle) \\ &= a^*UK|\psi\rangle + b^*UK|\phi\rangle \\ &= a^*T|\psi\rangle + b^*T|\phi\rangle. \end{aligned} \tag{8.98}$$

8.6.2 EVEN AND ODD OPERATORS

We say that an observable O is even or odd under time reversal according to whether it takes a positive or negative sign after the transformation

$$\boxed{TOT^{-1} = \pm O.}$$

(8.99)

For example, the momentum operator $\mathbf{p}$ is an odd operator and the position operator $\mathbf{r}$ is an even operator

$$T\mathbf{p}T^{-1} = -\mathbf{p}, \tag{8.100}$$
$$T\mathbf{r}T^{-1} = \mathbf{r}. \tag{8.101}$$

Applying the time-reversal operator to the fundamental commutator $[x, p_x] = i\hbar$, we obtain

$$Ti\hbar T^{-1} = T[x, p_x]T^{-1} = Txp_xT^{-1} - Tp_xxT^{-1} = -[x, p_x] = -i\hbar, \tag{8.102}$$

thus

$$TiT^{-1} = -i, \tag{8.103}$$

which is just the formula given in Eq.(8.94).

The orbital angular momentum $\mathbf{L}$ is a vector product of $\mathbf{r}$ and $\mathbf{p}$, $\mathbf{L} = \mathbf{r} \times \mathbf{p}$. Clearly, $\mathbf{L}$ is an odd operator. More generally, to preserve the commutation relation of the angular momentum

$$[J_\alpha, J_\beta] = i\hbar\varepsilon_{\alpha\beta\gamma}J_\gamma, \tag{8.104}$$

$\mathbf{J}$ must be an odd operator

$$T\mathbf{J}T^{-1} = -\mathbf{J} \tag{8.105}$$

since i changes sign under the time-reversal transformation This is a basic property of angular momentum under the time-reversal transformation.

For a particle of spin-1/2, the Pauli matrices changes under the time reversal as

$$T\sigma T^{-1} = -\sigma. \tag{8.106}$$

In this case, one can set

$$U = -i\sigma_y \tag{8.107}$$

up to an arbitrary phase factor. In this convention,

$$T = -i\sigma_y K. \tag{8.108}$$

8.6.3 NONDEGENERATE ENERGY EIGENSTATE

In a time-reversal symmetric system, if $|\Psi\rangle$ is an eigenstate of H with E the corresponding eigenvalue, so is $T|\Psi\rangle$ with the same eigenvalue

$$H(T|\Psi\rangle) = TH|\Psi\rangle = E(T|\Psi\rangle). \tag{8.109}$$

In real space, a quantum state can be expressed as

$$|\Psi\rangle = \int dx \Psi(x)|x\rangle, \tag{8.110}$$

where $\Psi(x)$ is the wave function. Applying the time-reversal operator to this state yields

$$T|\Psi\rangle = \int dx \Psi^*(x)|x\rangle, \tag{8.111}$$

hence the time-reversal transformation is simply to change the wave function by its complex conjugate

$$\Psi(x) \longrightarrow \Psi^*(x). \tag{8.112}$$

Using this property, it is simple to show

> **Theorem:** If the Hamiltonian is invariant under time reversal transformation and the energy eigenstate $|\Psi\rangle$ is nondegenerate, then the corresponding eigenfunction $\Psi(x)$ is real up to a x-independent phase factor.

8.6.4 KRAMERS DEGENERACY

If we do the time-reversal transformation twice, the state should return back to its original one, up to a phase factor. Hence, in general we should have

$$T^2|\Psi\rangle = e^{i\phi}|\Psi\rangle, \tag{8.113}$$

where ϕ is a phase. Using the property of antiunitary operator $T^2 = UKUK = UU^*$, the above equation leads to

$$UU^* = e^{i\phi}, \tag{8.114}$$

and

$$U = e^{i\phi}U^T \tag{8.115}$$

by multiplying both sides with U^T. Reiterating the above expression for U yields

$$U = e^{2i\phi}U, \tag{8.116}$$

hence $\exp(2i\phi) = 1$ and

$$\boxed{T^2 = e^{i\phi} = \pm 1.} \tag{8.117}$$

T^2 can take two values, 1 and -1. This is an important property of time-reversal operator, which differs from the parity operator whose square is always equal to one.

The case $T^2 = -1$ is highly nontrivial. It generally occurs in a half-integer spin system. For example, the time-reversal operator, given by Eq. (8.108), for the spin-1/2 system exhibits this property

$$T^2 = (-i\sigma_y K)^2 = -1. \tag{8.118}$$

In a time-reversal symmetric system, if $|\Psi\rangle$ is an eigenstate of the Hamiltonian with the eigenvalue E, $T|\Psi\rangle$ should also be an eigenstate with the same eigenvalue

$$HT|\Psi\rangle = TH|\Psi\rangle = ET|\Psi\rangle. \tag{8.119}$$

To see whether $T|\Psi\rangle$ is orthogonal to $|\Psi\rangle$ or not, let us evaluate the overlap between these two stats. Using the defining equation of the anti-unitary operator, we have

$$\langle \Psi | T\Psi \rangle = \langle T\Psi | T^2\Psi \rangle^* = \pm \langle T\Psi | \Psi \rangle^* = \pm \langle \Psi | T\Psi \rangle \tag{8.120}$$

for $T^2 = \pm 1$. In the case $T^2 = -1$, we further have

$$\langle \Psi | T\Psi \rangle = 0, \tag{8.121}$$

hence $T|\Psi\rangle$ is orthogonal to $|\Psi\rangle$, which means that the eigenenergy E is at least doubly degenerate. This double degeneracy, emerged in the time-reversal symmetric system with $T^2 = -1$, is called the Kramers degeneracy.

Eq. (8.120) does not impose any constraint on the degeneracy of an eigenenergy for the case $T^2 = 1$. However, if one of the eigenstates of the Hamiltonian is non-degenerate or with a $(2n + 1)$-fold degeneracy (n is integer) and the system is time-reversal symmetric, the Kramers degeneracy is absent and the above discussion implies that $T^2 = 1$.

8.7 PROBLEMS

1. Given a Hamiltonian

$$H = \frac{1}{2\mu} p_x^2 + V(x).$$

   a. Under which condition will it be translation invariant?
   b. If the system is translation invariant, show that the eigenvalues of H are two-fold degenerate, and that the degeneracy can be lifted out by considering the simultaneous eigenstates of H and p_x.

2. A particle moving in a three-dimensional potential $V(\mathbf{r})$ which depends only on y and z, show that the x-component of the momentum is a constant of motion.

3. Let $\Psi(x,t)$ be the wave function of a particle, show that $\Psi^*(x,-t)$ is the wave function of the particle with the momentum direction reversed.

4. Let $\langle p|\Psi\rangle = \Psi(p)$ be the wave function of $|\Psi\rangle$ in the momentum space. What is the momentum space wave function of the time-reversed state $T|\Psi\rangle$?

5. Let P be a parity operator, prove that if $[P,H] = 0$, a system that starts out in a state of certain parity maintain its parity.

6. An electron that is minimally coupled with an electromagnetic field is described by the Hamiltonian

$$H = \frac{1}{2\mu} (\mathbf{p} - e\mathbf{A})^2 - e\varphi.$$

   Show that a static magnetic field applied to the system breaks the time reversal invariance. How about a static electric field?

7. A particle of spin 1 is described by the Hamiltonian

$$H = aS_z^2 + b(S_x^2 - S_y^2),$$

   where (S_x, S_y, S_z) are the spin operators.

   a. Is this Hamiltonian invariant under the time reversal transformation?
   b. If the eigenstate $|jm\rangle$ of the spin operators $(\mathbf{S}^2, S_z)$ changes under the time-reversal transformation according to the formula

$$T|j,m\rangle = (-)^m |j,-m\rangle,$$

   where T is the time-reversal operator, how do the eigenstates of H change under the time-reversal transformation?

9 Approximate methods

9.1 GROUND STATE WAVE FUNCTION IS NODE FREE

From the examples discussed in Section 2.4 and Section 2.5, we notice that the wave function for the lowest bound state of a particle moving in a one-dimensional potential well is node free except at the boundary points where the potentials diverge. In fact, this observation holds very generally. It can be shown, using the argument that leads to the Perron-Frobenius theorem[1], that the phase of the ground state wave function of the Schrödinger equation in an arbitrary potential does not depend on the spatial coordinate, hence if we represent the ground state wave function

$$\Psi(\mathbf{r}) = R(\mathbf{r})e^{i\theta(\mathbf{r})/\hbar} \tag{9.1}$$

by its amplitude $R(\mathbf{r}) \geq 0$ and phase $\theta(\mathbf{r})$, then $\theta(\mathbf{r})$ should be $\mathbf{r}$-independent. This further implies that the normalizable ground state is a nonnegative real function

$$\boxed{\Psi(\mathbf{r}) = |\Psi(\mathbf{r})| = R(\mathbf{r}).} \tag{9.2}$$

To prove this, let us consider the difference between the energy of the particle in the state represented by the wave function $\Psi(\mathbf{r})$ and that represented by $|\Psi(\mathbf{r})|$

$$
\begin{aligned}
&E(\Psi) - E(|\Psi|) \\
&= \int d\mathbf{r}\Psi^* \left(-\frac{\hbar^2}{2\mu}\nabla^2 + V \right)\Psi - \int d\mathbf{r}R \left(-\frac{\hbar^2}{2\mu}\nabla^2 + V \right)R \\
&= -\frac{\hbar^2}{2\mu}\int d\mathbf{r}\left(\Psi^*\nabla^2\Psi - R\nabla^2 R \right) \\
&= -\frac{i\hbar}{2\mu}\int d\mathbf{r}\nabla \cdot (R^2\nabla\theta) + \frac{1}{2\mu}\int d\mathbf{r}R^2 (\nabla\theta)^2.
\end{aligned}
\tag{9.3}
$$

Using Eq. (1.32), the above equation can be further expressed as

$$E(\Psi) - E(|\Psi|) = -\frac{i\hbar}{2}\int d\mathbf{r}\nabla \cdot \mathbf{j} + \frac{1}{2\mu}\int d\mathbf{r}R^2 (\nabla\theta)^2, \tag{9.4}$$

where $\mathbf{j}$ is the probability current density. Since there is no net current flowing into or out of the system in the infinite boundary, we must have

$$\int d\mathbf{r}\nabla \cdot \mathbf{j} = \oint_{\Omega_\infty} d\mathbf{S} \cdot \mathbf{j} = 0. \tag{9.5}$$

---

[1] The Perron-Frobenius Theorem is a fundamental result for nonnegative matrices. It states that a real square matrix with positive entries has a unique largest real eigenvalue and that the corresponding eigenvector can be set to have positive components only. This theorem can be readily extended to a real symmetric square matrix whose off-diagonal matrix elements are all negative and show that its ground state is unique and all the components of the corresponding eigenvector can be set strictly positive.

DOI: 10.1201/9781003174882-9

169

Thus the energy difference is

$$E(\Psi) - E(|\Psi|) = \frac{1}{2\mu} \int d\mathbf{r} R^2 (\nabla \theta)^2 \geq 0, \tag{9.6}$$

which shows that the phase of the wave function, θ, is spatial independent except in the region where $R = 0$, hence Eq. (9.2) is valid. It further implies that there is no node across which the ground state wave function changes sign.

9.2 VARIATIONAL ANSATZ

In case you want calculate the ground state energy without solving the eigen-equation of the Hamiltonian, you can make a guess on the ground state wave function with some unknown parameters, and determine these parameters by minimizing the energy expectation value of the trial wave function based on the variational principle.

Variational principle:

For any normalized wave function Ψ, its energy expectation is always higher than or equal to the true ground state energy E_0, namely

$$E_0 \leq \langle \Psi | H | \Psi \rangle. \tag{9.7}$$

The equality holds if and only if $|\Psi\rangle$ is one of the ground states.

Inequality (9.7) constitutes the basis of the variational method[2] for the approximate calculation of the ground state energy and the wave function. It consists of evaluating the expectation value of the Hamiltonian using a trial wave function Ψ which contains a number of variational parameters. These parameters are determined by minimizing the energy of this trial state. The minimized energy such obtained, as revealed by the variational principle, is never lower than the true ground state energy. It provides an upper bound for the ground state energy.

Apparently, a successful implementation of the variational principle relies highly on the choice of the trial wave function. In principle, a good trial wave function should be reasonably simple to evaluate and yet contain sufficiently many features of the exact ground state. Some of the features can be deduced from symmetry arguments and from the node structure of the wave function.

To prove the above inequality, let us expand $|\Psi\rangle$ using the eigenstates of the Hamiltonian

$$|\Psi\rangle = \sum_n C_n |\Psi_n\rangle, \tag{9.8}$$

---

[2]The variational method is a quantum generalization of the Rayleigh-Ritz method used for finding approximations to eigenvalue equations that are difficult to solve exactly. The method was invented by Walther Ritz in 1909, but it bears some similarity to the Rayleigh quotient. It is also called the Rayleigh-Ritz method.

where $|\Psi_n\rangle$ is the normalized eigenstate of H

$$H|\Psi_n\rangle = E_n|\Psi_n\rangle, \tag{9.9}$$

and C_n is the wave function of $|\Psi\rangle$ in this basis representation.

The expectation value of H for the state $|\Psi\rangle$ is

$$\langle\Psi|H|\Psi\rangle = \sum_{mn} C_m^* C_n \langle\Psi_m|H|\Psi_n\rangle = \sum_n E_n|C_n|^2 \geq E_0 \sum_n |C_n|^2. \tag{9.10}$$

Since $|\Psi\rangle$ is normalized, we have

$$1 = \langle\Psi|\Psi\rangle = \sum_n |C_n|^2. \tag{9.11}$$

From Eqs. (9.10) and (9.11), we immediately obtain the inequality (9.7).

9.2.1 HALF-HARMONIC OSCILLATOR

As an example, we take the half-harmonic oscillator, defined by the potential

$$V(x) = \begin{cases} \dfrac{1}{2}\mu\omega^2 x^2 & x \geq 0, \\[2mm] \infty & x < 0, \end{cases} \tag{9.12}$$

to demonstrate how to calculate the ground state energy using the variational principle. In the positive x part, it is just the ordinary harmonic oscillator with the Hamiltonian discussed in Section 4.1,

$$H = -\frac{\hbar^2}{2\mu}\frac{d^2}{dx^2} + \frac{1}{2}\mu\omega^2 x^2, \qquad x \geq 0. \tag{9.13}$$

However, in the negative axis, the potential becomes infinite so that the wave function $\Psi(x)$ vanishes. From the continuity of the wave function, we require that $\Psi(0) = 0$.

This half-harmonic oscillator Hamiltonian can be diagonalized using the same approach as for solving the ordinary harmonic oscillator. It is simple to show that all the eigenstates of the ordinary harmonic oscillator satisfy the Schrödinger eigenequation in the positive axis. However, only the odd parity solutions satisfy the condition $\Psi(0) = 0$. Therefore, the eigenenergy of the half-harmonic oscillator is given by the odd-parity eigenenergy of the ordinary harmonic oscillator, i.e.

$$E_n = \left(n + \frac{1}{2}\right)\hbar\omega, \qquad n = 1, 3, 5, \ldots \tag{9.14}$$

In the ground state, $n = 1$ and $E_1 = 3\hbar\omega/2$.

Now let us use the variational principle to find an "approximate" result for the ground state energy. As the ground state wave function is node free except at the potential edge, we assume the trial wave function $\Psi(x)$ to be

$$\Psi(x) = Axe^{-\alpha x^2}, \tag{9.15}$$

where α is the variational parameter. This wave function vanishes at $x = 0$, satisfying the continuous condition of wave function at the boundary site. A is the normalization constant, determined by

$$1 = \int_0^\infty dx |\Psi(x)|^2 = A^2 \int_0^\infty dx x^2 e^{-2\alpha x^2} = \frac{1}{8\alpha^{3/2}} \sqrt{\frac{\pi}{2}} A^2, \quad (9.16)$$

so that

$$A = \left(\frac{128\alpha^3}{\pi} \right)^{1/4}. \quad (9.17)$$

The expectation values of the kinetic and potential energies are given respectively by

$$
\begin{aligned}
\langle T \rangle &= \langle \Psi | -\frac{\hbar^2}{2\mu} \frac{d^2}{dx^2} | \Psi \rangle \\
&= -\frac{\hbar^2}{2\mu} A^2 \int_0^\infty dx x e^{-\alpha x^2} \frac{d^2}{dx^2} \left(x e^{-\alpha x^2} \right) \\
&= \frac{\hbar^2}{2\mu} A^2 \int_0^\infty dx x^2 e^{-2\alpha x^2} (6\alpha - 4\alpha^2 x^2) \\
&= \frac{3\hbar^2 \alpha}{2\mu}, \quad (9.18)
\end{aligned}
$$

and

$$
\begin{aligned}
\langle V \rangle &= \langle \Psi | \frac{1}{2} \mu \omega^2 x^2 | \Psi \rangle \\
&= \frac{1}{2} \mu \omega^2 A^2 \int_0^\infty dx x^4 e^{-2\alpha x^2} \\
&= \frac{3\mu \omega^2}{8\alpha}. \quad (9.19)
\end{aligned}
$$

Therefore, the total energy is

$$\langle H \rangle = \langle T \rangle + \langle V \rangle = \frac{3\hbar^2 \alpha}{2\mu} + \frac{3\mu \omega^2}{8\alpha}. \quad (9.20)$$

By minimizing the energy with respect to a, we obtain the equation

$$\frac{d}{d\alpha} \langle H \rangle = \frac{3\hbar^2}{2\mu} - \frac{3\mu \omega^2}{8\alpha^2} = 0. \quad (9.21)$$

α is then found to be $\alpha = \mu \omega / (2\hbar)$. Substituting it into $\langle H \rangle$, the variational energy is found to be

$$\langle H \rangle_v = \frac{3}{2} \hbar \omega. \quad (9.22)$$

This result sets up an upper bound of the ground state energy. It happens to be the exact ground state energy of the half-harmonic oscillator, because the trial wave function "happens" to have precisely the form of the actual ground state.

9.2.2 GROUND STATE OF HELIUM

We now use the variational principle to estimate the ground state energy of the helium atom. A helium consists of two electrons and two protons, and is described by the Hamiltonian

$$H = h_1 + h_2 + V_{12}, \tag{9.23}$$

where

$$h_i = -\frac{\hbar^2}{2\mu}\nabla_i^2 - \frac{2e^2}{4\pi\varepsilon_0 r_i} \tag{9.24}$$

is the Hamiltonian of the i'th electron, and

$$V_{12} = \frac{e^2}{4\pi\varepsilon_0 |\mathbf{r}_1 - \mathbf{r}_2|}$$

is the Coulomb repulsion between two electrons. There is no analytic solution to this Hamiltonian, although it looks simple. Nevertheless, the ground state energy of helium has been accurately measured in the laboratory,

$$E_{ex} = -78.975\text{eV}. \tag{9.25}$$

It sets a reference value for comparison with the variational calculation.

The Hamiltonian does not depend on the spin degrees of freedom, so the total spin of the two electrons is a constant of motion. This implies that the wave function can be decoupled into a product of a spin wave function and a spatial one $\Psi(\mathbf{r}_1, \mathbf{r}_2)$, and the spin state should be either singlet or triplet. If the two electrons form a spin singlet, the spatial wave function should be symmetric with respect to the exchange of $\mathbf{r}_1$ and $\mathbf{r}_2$. On the other hand, if the two electrons form a spin triplet state, the spatial wave function should be antisymmetric. For the ground state, it turns out that the spin singlet state has a lower energy, so we need only consider the case $\Psi(\mathbf{r}_1, \mathbf{r}_2)$ is symmetric.

A crude but natural guess for the ground state wave function is to assume that both electrons are in the ground state of the $Z = 2$ hydrogenic atom, i.e.

$$\Psi(\mathbf{r}_1, \mathbf{r}_2) = \Psi_{100}(\mathbf{r}_1)\Psi_{100}(\mathbf{r}_2) = \frac{Z^3}{\pi a_0^3}e^{-Z(r_1+r_2)/a_0} = \frac{1}{\pi\alpha^3}e^{-(r_1+r_2)/\alpha}, \tag{9.26}$$

where $\alpha = a_0/Z$ and a_0 is the Bohr radius (we ignore the difference between the mass of electron and its reduced mass), and

$$\Psi_{100}(\mathbf{r}) = \frac{1}{\sqrt{\pi}\alpha^{3/2}}e^{-r/\alpha} \tag{9.27}$$

is the ground state wave function of the hydrogenic atom. $\Psi(\mathbf{r}_1, \mathbf{r}_2)$ such defined is actually the exact solution of the Hamiltonian without the V_{12} term.

However, the V_{12} term is not negligible. In fact, it will generate a large repulsive energy to rise the total ground state energy. We need to modify the above wave function to reduce this repulsive Coulomb energy. An inherent feature of a many-electron system that should be considered is the screening effect of the negative charge cloud of an electron, which can effectively reduce the positive charge of the nucleon that acts on the other electron. This may reduce the repulsive interaction between the two electrons. It suggests that the effective Z should be less than 2, and we should take Z or α as a free parameter and determine it by minimizing the ground state energy.

Taking Ψ defined in Eq. (9.26) as a trial wave function with α the variational parameter, the expectation value of H is a sum of two terms

$$\langle H \rangle = \langle \Psi | h_1 + h_2 + V_{12} | \Psi \rangle = 2 \langle \Psi | h_1 | \Psi \rangle + \langle \Psi | V_{12} | \Psi \rangle, \tag{9.28}$$

where $\langle \Psi | h_1 | \Psi \rangle$ is the energy of the first electron

$$
\begin{aligned}
\langle \Psi | h_1 | \Psi \rangle &= \frac{1}{\pi^2 \alpha^6} \int d\mathbf{r}_1 d\mathbf{r}_2 e^{-(r_1+r_2)/\alpha} h_1 e^{-(r_1+r_2)/\alpha} \\
&= \frac{1}{\pi \alpha^3} \int d\mathbf{r}_1 e^{-r_1/\alpha} \left(-\frac{\hbar^2}{2\mu} \nabla_1^2 - \frac{2e^2}{4\pi\varepsilon_0 r_1} \right) e^{-r_1/\alpha} \\
&= \frac{4}{\alpha^3} \int dr r^2 \left(-\frac{\hbar^2}{2\mu r^2} e^{-r/\alpha} \frac{\partial}{\partial r} r^2 \frac{\partial}{\partial r} e^{-r/\alpha} - \frac{2e^2}{4\pi\varepsilon_0 r} e^{-2r/\alpha} \right) \\
&= \frac{\hbar^2}{2\mu\alpha^2} - \frac{e^2}{2\pi\varepsilon_0 \alpha}.
\end{aligned} \tag{9.29}
$$

The expectation value of V_{12} is given by the integral

$$
\begin{aligned}
\langle \Psi | V_{12} | \Psi \rangle &= \frac{e^2}{4\varepsilon_0 \pi^3 \alpha^6} \int d\mathbf{r}_1 d\mathbf{r}_2 \frac{1}{|\mathbf{r}_1 - \mathbf{r}_2|} e^{-2(r_1+r_2)/\alpha} \\
&= \frac{e^2}{4\varepsilon_0 \pi^3 \alpha^6} \int d\mathbf{r}_1 e^{-2r_1/\alpha} A(\mathbf{r}_1).
\end{aligned} \tag{9.30}
$$

where

$$A(\mathbf{r}_1) = \int d\mathbf{r}_2 \frac{1}{|\mathbf{r}_1 - \mathbf{r}_2|} e^{-2r_2/\alpha}. \tag{9.31}$$

Assuming θ to be the angle between $\mathbf{r}_1$ and $\mathbf{r}_2$, the above integral can be expressed as

$$A(\mathbf{r}_1) = \int_0^\infty 2\pi r_2^2 e^{-2r_2/\alpha} dr_2 \int_0^\pi \frac{\sin\theta d\theta}{\sqrt{r_1^2 + r_2^2 - 2r_1 r_2 \cos\theta}}. \tag{9.32}$$

The θ integral is

$$\int_0^\pi \frac{\sin\theta d\theta}{\sqrt{r_1^2 + r_2^2 - 2r_1 r_2 \cos\theta}} = \left. \frac{\sqrt{r_1^2 + r_2^2 - 2r_1 r_2 \cos\theta}}{r_1 r_2} \right|_0^\pi = \frac{2}{\max(r_1, r_2)}. \tag{9.33}$$

It follows that

$$
\begin{aligned}
A(\mathbf{r}_1) &= \int_0^{r_1} \frac{4\pi r_2^2 e^{-2r_2/\alpha}}{r_1} dr_2 + \int_{r_1}^\infty 4\pi r_2 e^{-2r_2/\alpha} dr_2 \\
&= \frac{\pi\alpha^3}{r_1} \left[1 - \left(1 + \frac{r_1}{\alpha} \right) e^{-2r_1/\alpha} \right],
\end{aligned} \tag{9.34}
$$

and the expectation value of V_{12} is

$$
\begin{aligned}
\langle\Psi|V_{12}|\Psi\rangle &= \frac{e^2}{4\varepsilon_0\pi^2\alpha^3}\int d\mathbf{r}_1 e^{-2r_1/\alpha}\frac{1}{r_1}\left[1-\left(1+\frac{r_1}{\alpha}\right)e^{-2r_1/\alpha}\right] \\
&= \frac{5e^2}{32\varepsilon_0\pi\alpha}
\end{aligned}
\tag{9.35}
$$

Therefore, the total energy is

$$
\langle H\rangle = 2\langle\Psi|h_1|\Psi\rangle + \langle\Psi|V_{12}|\Psi\rangle = \frac{\hbar^2}{\mu\alpha^2} - \frac{27e^2}{32\pi\varepsilon_0\alpha}
\tag{9.36}
$$

Below we compare the results obtained with or without considering the screening effect:

1. Without considering the screening effect, $\alpha = a_0/Z$ and $Z = 2$. The corresponding energy is

$$
\langle H\rangle = \frac{\hbar^2 Z^2}{\mu a_0^2} - \frac{27e^2 Z}{32\pi\varepsilon_0 a_0} = \frac{11}{2}E_0 = -74.8 \text{ eV},
\tag{9.37}
$$

where $E_0 = -e^2/(8\pi\varepsilon_0 a_0) = -13.6eV$ is the ground state energy of the hydrogen atom. In comparison with the experimental result, the relative error of this energy is

$$
\frac{\langle H\rangle - E_{ex}}{E_{ex}} = 5.3\%.
\tag{9.38}
$$

The error is a bit large, but not too bad, considering that there is no adjustable parameter in the trial wave function.

2. Considering the screening effect, we determine α by variationally minimizing the ground state energy. This yields

$$
\alpha = \frac{64\pi\varepsilon_0\hbar^2}{27\mu e^2} = \frac{16}{27}a_0.
\tag{9.39}
$$

The effective Z is $Z_{eff} = 27/16$, which is indeed less than 2. The minimal energy is

$$
\langle H\rangle = -\frac{729}{128}\frac{e^2}{8\pi\varepsilon_0 a_0} = -\frac{729}{128}|E_0| \approx -77.5 \text{ eV}.
\tag{9.40}
$$

Now the relative error is

$$
\frac{\langle H\rangle - E_{ex}}{E_{ex}} = 1.9\%.
\tag{9.41}
$$

As expected, the variational calculation produces a better estimation for the ground state energy. There is no doubt that more accurate ground state energy could be obtained if more complicated trial wave functions, with more adjustable parameters, are adopted.

9.3 STATIONARY PERTURBATION THEORY

Besides the variational approach, perturbation theory is another approximate method that is more commonly used for solving the Schrödinger equation in which part of the Hamiltonian is exactly soluble and the correction imposed by the other part of the Hamiltonian is small. It can be used to study not just the ground state, but also all excitation states. This kind of approximate methods can be divided into two groups, according to whether the perturbation is time-independent or time-dependent. Here we discuss the Rayleigh-Schrödinger perturbation theory, which consider how the discrete energy levels and the corresponding eigenfunctions are corrected when a perturbation is applied.

Perturbation theory starts from a Hamiltonian of the form

$$H = H_0 + \lambda H_1, \tag{9.42}$$

where H does not differ much from H_0. H_0 is called the unperturbed Hamiltonian. λH_1 is the perturbation, which has to be small. λ is a real parameter which is used to label the power of perturbation in the series expansion of H_1. After the perturbation calculation, we set $\lambda = 1$.

We assume that H_0 is rigorously soluble and its eigenvalues and eigenfunctions are known

$$H_0 \left| \Psi_n^0 \right\rangle = E_n^0 \left| \Psi_n^0 \right\rangle, \tag{9.43}$$

where E_n^0 is the n'th eigenvalue of H_0 and $\left| \Psi_n^0 \right\rangle$ the corresponding eigenvector, which is orthonormalized

$$\left\langle \Psi_{n'}^0 | \Psi_n^0 \right\rangle = \delta_{nn'}. \tag{9.44}$$

The eigenvalue problem we want to solve is

$$(H_0 + \lambda H_1) \left| \Psi_n \right\rangle = E_n \left| \Psi_n \right\rangle. \tag{9.45}$$

We assume E_n is adiabatically connected to E_n^0 by varying continuously λ from 0 to 1 so that we can expand both the eigenvalue and the eigenfunction in the power of λ

$$\begin{align} E_n &= E_n^0 + \lambda E_n^1 + \lambda^2 E_n^2 + \cdots, \tag{9.46} \\ \left| \Psi_n \right\rangle &= \left| \Psi_n^0 \right\rangle + \lambda \left| \Psi_n^1 \right\rangle + \lambda^2 \left| \Psi_n^2 \right\rangle + \cdots. \tag{9.47} \end{align}$$

Substituting these equations into Eq. (9.45), we obtain

$$\begin{align} &(H_0 + \lambda H_1) \left(\left| \Psi_n^0 \right\rangle + \lambda \left| \Psi_n^1 \right\rangle + \lambda^2 \left| \Psi_n^2 \right\rangle + \cdots \right) \\ = &\left(E_n^0 + \lambda E_n^1 + \lambda^2 E_n^2 + \cdots \right) \left(\left| \Psi_n^0 \right\rangle + \lambda \left| \Psi_n^1 \right\rangle + \lambda^2 \left| \Psi_n^2 \right\rangle + \right). \tag{9.48} \end{align}$$

We now equate the coefficients of equal powers of λ on both sides of the equation.

To the zeroth order of λ, we obtain Eq. (9.43) as expected. More generally, by equating the coefficients of λ^α ($\alpha > 0$), we have

$$\boxed{H_0 \left| \Psi_n^\alpha \right\rangle + H_1 \left| \Psi_n^{\alpha-1} \right\rangle = E_n^0 \left| \Psi_n^\alpha \right\rangle + E_n^1 \left| \Psi_n^{\alpha-1} \right\rangle + \cdots + E_n^\alpha \left| \Psi_n^0 \right\rangle.} \tag{9.49}$$

9.3.1 FIRST ORDER CORRECTION

To the first order approximation, Eq. (9.49) becomes

$$H_0\left|\Psi_n^1\right\rangle + H_1\left|\Psi_n^0\right\rangle = E_n^0\left|\Psi_n^1\right\rangle + E_n^1\left|\Psi_n^0\right\rangle. \tag{9.50}$$

To multiply both sides of the equation by $\left\langle\Psi_n^0\right|$, we obtain

$$\left\langle\Psi_n^0|H_0\left|\Psi_n^1\right\rangle + \left\langle\Psi_n^0\right|H_1\left|\Psi_n^0\right\rangle = E_n^0\langle\Psi_n^0\left|\Psi_n^1\right\rangle + E_n^1\langle\Psi_n^0\left|\Psi_n^0\right\rangle. \tag{9.51}$$

The first term on the left hand side cancels the third term. Therefore, the first order correction to the energy is

$$\boxed{E_n^1 = \left\langle\Psi_n^0\right| H_1\left|\Psi_n^0\right\rangle.} \tag{9.52}$$

It indicates that the leading order correction to the energy is just the perturbation Hamiltonian H_1 averaged over the corresponding unperturbed state.

Up to the first order correction, the ground state energy is

$$E_0 = E_0^0 + E_0^1 = \langle\Psi_0^0|H_0 + H_1|\Psi_0^0\rangle. \tag{9.53}$$

This is equivalent to taking $|\Psi_0^0\rangle$ as a variational wave function for the ground state of H. Clearly, it satisfies the variational principle, and E_0 such obtained is an upper bound to the true ground state energy.

To find the first order correction to the wave function, we rewrite Eq. (9.50) as

$$\left(H_0 - E_n^0\right)\left|\Psi_n^1\right\rangle = \left(E_n^1 - H_1\right)\left|\Psi_n^0\right\rangle. \tag{9.54}$$

This gives

$$\begin{aligned}
\left|\Psi_n^1\right\rangle &= \frac{1}{H_0 - E_n^0}\left(E_n^1 - H_1\right)\left|\Psi_n^0\right\rangle \\
&= \sum_m \frac{1}{H_0 - E_n^0}|\Psi_m^0\rangle\langle\Psi_m^0|E_n^1 - H_1|\Psi_n^0\rangle \\
&= \sum_{m\neq n} \frac{1}{H_0 - E_n^0}|\Psi_m^0\rangle\langle\Psi_m^0|E_n^1 - H_1|\Psi_n^0\rangle \\
&= -\sum_{m\neq n} \frac{1}{E_m^0 - E_n^0}|\Psi_m^0\rangle\langle\Psi_m^0|H_1|\Psi_n^0\rangle, \tag{9.55}
\end{aligned}$$

so that

$$\boxed{\left|\Psi_n^1\right\rangle = \sum_{m\neq n} \frac{\left\langle\Psi_m^0\right| H_1\left|\Psi_n^0\right\rangle}{E_n^0 - E_m^0}\left|\Psi_m^0\right\rangle.} \tag{9.56}$$

9.3.2 SECOND ORDER CORRECTION

To the second order in λ^2, Eq. (9.49) becomes

$$H_0\left|\Psi_n^2\right\rangle + H_1\left|\Psi_n^1\right\rangle = E_n^0\left|\Psi_n^2\right\rangle + E_n^1\left|\Psi_n^1\right\rangle + E_n^2\left|\Psi_n^0\right\rangle. \tag{9.57}$$

Again by multiplying the both sides of the equation with $\left\langle\Psi_n^0\right|$, we obtain

$$\begin{aligned} E_n^2 &= \left\langle\Psi_n^0\middle|H_0 - E_n^0\middle|\Psi_n^2\right\rangle + \left\langle\Psi_n^0\middle|H_1 - E_n^1\middle|\Psi_n^1\right\rangle \\ &= \left\langle\Psi_n^0\middle|H_1 - E_n^1\middle|\Psi_n^1\right\rangle. \end{aligned} \tag{9.58}$$

Substituting Eq. (9.56) into the above equation yields

$$\boxed{E_n^2 = \sum_{m\neq n} \frac{\left|\left\langle\Psi_m^0\middle|H_1\middle|\Psi_n^0\right\rangle\right|^2}{E_n^0 - E_m^0}.} \tag{9.59}$$

The second order correction to the energy can therefore be obtained by summing over all the state $\left|\Psi_m^0\right\rangle$ with $m \neq n$. If n is the ground state, then $E_m^0 - E_n^0$ is always larger than 0, and $E_n^2 < 0$. So the second order correction is always to lower the ground state energy.

The correction to the wave function is given by

$$(H_0 - E_n^0)\left|\Psi_n^2\right\rangle = (E_n^1 - H_1)\left|\Psi_n^1\right\rangle + E_n^2\left|\Psi_n^0\right\rangle. \tag{9.60}$$

Using Eq. (9.58), it can be further expressed as

$$(H_0 - E_n^0)\left|\Psi_n^2\right\rangle = \sum_{m\neq n}\left|\Psi_m^0\right\rangle\left\langle\Psi_m^0\middle|E_n^1 - H_1\middle|\Psi_n^1\right\rangle. \tag{9.61}$$

We therefore have

$$\boxed{\left|\Psi_n^2\right\rangle = \sum_{m\neq n}\frac{\left\langle\Psi_m^0\middle|H_1 - E_n^1\middle|\Psi_n^1\right\rangle}{E_n^0 - E_m^0}\left|\Psi_m^0\right\rangle.} \tag{9.62}$$

9.3.3 ANHARMONIC OSCILLATOR

We now use the stationary perturbation theory to calculate the corrections to the energy levels and wave functions of the harmonic oscillator perturbed by an anharmonic potential that is proportional to x^3. The potential energy becomes

$$V(x) = \frac{1}{2}\mu\omega^2 x^2 + \alpha\left(\frac{2\mu\omega}{\hbar}\right)^{3/2}\hbar\omega x^3, \tag{9.63}$$

where α is a dimensionless parameter that measures the strength of the anharmonic potential. We assume that the second term in $V(x)$ is small in comparison with the first term, so that

$$H_0 = -\frac{\hbar^2}{2\mu}\frac{d^2}{dx^2} + \frac{1}{2}\mu\omega^2 x^2 \tag{9.64}$$

can be taken as the unperturbed Hamiltonian and

$$H_1 = \alpha \left(\frac{2\mu\omega}{\hbar}\right)^{3/2} \hbar\omega x^3 \tag{9.65}$$

as the perturbation. H_0 is exactly soluble and a thorough discussion on its solution is given in Section 4.1.

Using the ladder operators defined in 4.1.1, x can be expressed as

$$x = \sqrt{\frac{\hbar}{2\mu\omega}}(a + a^\dagger) \tag{9.66}$$

and the perturbation Hamiltonian becomes

$$H_1 = \alpha\hbar\omega(a + a^\dagger)^3. \tag{9.67}$$

By acting $(a + a^\dagger)^3$ on the eigenstate $|n\rangle$ of the harmonic oscillator H_0, we find that

$$
\begin{aligned}
(a + a^\dagger)^3 |n\rangle &= \sqrt{(n+1)(n+2)(n+3)}|n+3\rangle + 3(n+1)^{3/2}|n+1\rangle \\
&\quad + 3n^{3/2}|n-1\rangle + \sqrt{n(n-1)(n-2)}|n-3\rangle.
\end{aligned} \tag{9.68}
$$

Since

$$\langle n|(a + a^\dagger)^3|n\rangle = 0, \tag{9.69}$$

the first order correction to the eigenenergy vanishes

$$E_n^1 = \langle n|H_1|n\rangle = 0. \tag{9.70}$$

However, the first order corrections to the eigenfunctions and the second order corrections to the eigenlevels do not vanish.

From Eq. (9.68), the nonzero matrix elements of H_1 are found to be

$$\langle n+3|H_1|n\rangle = \sqrt{(n+1)(n+2)(n+3)}\,\alpha\hbar\omega, \tag{9.71}$$
$$\langle n+1|H_1|n\rangle = 3(n+1)^{3/2}\alpha\hbar\omega, \tag{9.72}$$
$$\langle n-1|H_1|n\rangle = 3n^{3/2}\alpha\hbar\omega, \tag{9.73}$$
$$\langle n-3|H_1|n\rangle = \sqrt{n(n-1)(n-2)}\,\alpha\hbar\omega. \tag{9.74}$$

The parity of the eigenstate is changed by H_1 — if $|n\rangle$ is parity even, then the final states are parity odd, and vice versa. This is because x^3 is an odd parity operator.

The second order correction to the eigevlevel is therefore given by

$$E_n^2 = \frac{|\langle n+3|H_1|n\rangle|^2}{-3\hbar\omega} + \frac{|\langle n+1|H_1|n\rangle|^2}{-\hbar\omega} + \frac{|\langle n-1|H_1|n\rangle|^2}{\hbar\omega} + \frac{|\langle n-3|H_1|n\rangle|^2}{3\hbar\omega}.$$

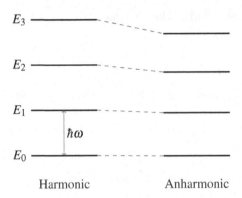

Figure 9.1 Energy level spacing changes induced by the anharmonic interaction. Here $60\alpha^2 = 0.05$ is used for illustration.

Substituting the expressions of the matrix elements into the above equation, we find that

$$
\begin{aligned}
E_n^2 &= \left[\frac{|(n+1)(n+2)(n+3)}{-3} - 9(n+1)^3 + 9n^3 + \frac{n(n-1)(n-2)}{3}\right]\alpha^2\hbar\omega \\
&= \left[-9(3n^2+3n+1) - (3n^2-3n+2)\right]\alpha^2\hbar\omega \\
&= -(30n^2+30n+11)\alpha^2\hbar\omega
\end{aligned}
$$
(9.75)

The second order corrections to the first three eigenlevels are $(-11, -71, -191)\alpha^2\hbar\omega$. Thus, to second order in perturbation, the energy of the anharmonic oscillator is

$$
E_n \approx \left(n+\frac{1}{2}\right)\hbar\omega - (30n^2+30n+11)\alpha^2\hbar\omega.
$$
(9.76)

Clearly, the energy levels are no longer equally separated. The energy difference between two neighboring energy levels is

$$
E_{n+1} - E_n = \left[1 - 60(n+1)\alpha^2\right]\hbar\omega.
$$
(9.77)

It decreases with increasing n. A schematic plot of the level spacing before and after the perturbation is shown in Fig. 9.1. The correction increases rapidly with n. Apparently, the perturbation holds only for low-energy eigenstates.

Similarly, we can find the first order correction to the eigenfunction

$$
\begin{aligned}
|\Psi_n^1\rangle &= \frac{\langle n+3|H_1|n\rangle}{-3\hbar\omega}|n+3\rangle + \frac{\langle n+1|H_1|n\rangle}{-\hbar\omega}|n+1\rangle \\
&\quad + \frac{\langle n-1|H_1|n\rangle}{\hbar\omega}|n-1\rangle + \frac{\langle n-3|H_1|n\rangle}{3\hbar\omega}|n-3\rangle \\
&= -\frac{\alpha}{3}\sqrt{(n+1)(n+2)(n+3)}|n+3\rangle - 3\alpha(n+1)^{3/2}|n+1\rangle \\
&\quad + 3\alpha n^{3/2}|n-1\rangle + \frac{\alpha}{3}\sqrt{n(n-1)(n-2)}|n-3\rangle.
\end{aligned}
$$
(9.78)

It is straightforward but a little bit tedious to write down the second order correction to the eigenfunction. Without calculation, however, we know that $|\Psi_n^2\rangle$ has the same parity as $|n\rangle$.

9.4   DEGENERATE PERTURBATION THEORY

The derivation given in the preceding section holds when the unperturbed energy level E_n^0 is nondegenerate. We now consider the correction of the perturbation when the unperturbed energy level is degenerate.

If E_n^0 is degenerate, there are several unperturbed basis states $|\Psi_{nr}^0\rangle$ $(r = 1, \cdots, \alpha)$ correspond to this level. We assume all these basis states are orthonormalized

$$\langle \Psi_{mr}^0 | \Psi_{ns}^0 \rangle = \delta_{mn} \delta_{rs}. \tag{9.79}$$

Let us introduce a set of re-orthonormalized zeroth order basis states $|\chi_{nr}^0\rangle$ which yield the first term in the expansion of the exact basis states $|\Psi_{nr}\rangle$ in power of λ

$$|\Psi_{nr}\rangle = |\chi_{nr}^0\rangle + \lambda |\Psi_{nr}^1\rangle + \lambda^2 |\Psi_{nr}^2\rangle + \cdots. \tag{9.80}$$

$|\chi_{nr}^0\rangle$ are linear combinations of the unperturbed basis states

$$|\chi_{nr}^0\rangle = \sum_s C_{rs} |\Psi_{ns}^0\rangle. \tag{9.81}$$

The coefficients C_{rs} will be determined. Correspondingly, the correction to the energy can be written as

$$E_{nr} = E_n^0 + \lambda E_{nr}^1 + \lambda^2 E_{nr}^2 + \cdots. \tag{9.82}$$

Substituting Eqs. (9.80) and (9.82) into the Schrodinger equation

$$(H_0 + \lambda H_1)|\Psi_{nr}\rangle = E_{nr}|\Psi_{nr}\rangle \tag{9.83}$$

and equating the coefficients of λ, we find that

$$H_0 |\Psi_{nr}^1\rangle + H_1 |\chi_{nr}^0\rangle = E_n^0 |\Psi_{nr}^1\rangle + E_{nr}^1 |\chi_{nr}^0\rangle. \tag{9.84}$$

Multiplying both side by $\langle \Psi_{ns}^0 |$, we obtain

$$\sum_{s'} \left[\langle \Psi_{ns}^0 | H_1 | \Psi_{ns'}^0 \rangle - E_{nr}^1 \delta_{ss'} \right] C_{rs'} = 0. \tag{9.85}$$

This is a linear system of equations. A nontrivial solution is obtained if the determinant of the quantity in the brackets vanishes, i.e.

$$\boxed{\det \left| \langle \Psi_{ns}^0 | H_1 | \Psi_{ns'}^0 \rangle - E_{nr}^1 \delta_{ss'} \right| = 0.} \tag{9.86}$$

It has α real roots. If all these roots are distinct, the degeneracy is removed to first order in perturbation. On the other hand, if some roots are identical, the degeneracy is only partially removed. This may happen when H_0 and H_1 share some symmetries.

By solving the above secular equation, one can also determine the coefficients C_{rs}, from which one can further determine the first order correction to the wave function $|\Psi_{nr}^1\rangle$ in a way similar to that introduced in Section 9.3.

9.4.1 LINEAR STARK EFFECT

The Stark effect, which was first discovered by the German physicist Johannes Stark in 1913, is the shifting and splitting of spectral lines of atoms and molecules due to presence of an external electric field. It is the electric analogue of the Zeeman effect where a spectral line is split into several components due to the presence of a magnetic field. The amount of splitting or shifting is called the Stark splitting or Stark shift.

The linear Stark effect describes the level splitting of a hydrogen atom in a homogeneous electric field. The splitting of the energy levels of other atoms is proportional to the second power in the field strength, which is a quadratic Stark effect.

The Stark splitting is a quantum effect. It is impossible to understand this effect in classical theory. Experimentally, the applied electric field is about $10^4 - 10^5$ V/cm, much smaller than the internal electric field generated by the nucleus, which is of the order $e/a_0^2 \sim 5 \times 10^9$ V/cm (a_0 is the Bohr radius of hydrogen atom). Thus we can treat the applied electric field as a perturbation.

We do not need to consider the first energy level of the hydrogen atom, i.e. the level with the principal quantum number $n = 1$, because this energy level is non-degenerate. Here we only consider the splitting of the second level ($n = 2$) under an applied electric field. There are four degenerate states at the $n = 2$ level. Among them, one is in the $l = 0$ state, $|nlm\rangle = |200\rangle$, and three are in the $l = 1$ states, $(|210\rangle, |211\rangle, |21\bar{1}\rangle)$.

We assume the electric field is applied along the z-axis. The potential energy corresponding to this field is given by

$$H_1 = -e\mathbf{r} \cdot \mathbf{E}_s = -eE_s r \cos \theta = -eE_s z, \tag{9.87}$$

where θ is the angle between $\mathbf{r}$ and the Stark field $\mathbf{E}_s$. As H_1 commutes with L_z, thus the eigenvalue of L_z is not altered by H_1, hence

$$\langle 2l'm' | H_1 | 2lm \rangle = 0 \quad \text{if } m \neq m'. \tag{9.88}$$

Moreover, as H_1 is parity odd under the spatial reflection with respect to the xy plane, it can be shown that

$$\langle 2lm | \cos \theta | 2lm \rangle = 0 \tag{9.89}$$

for all the four states.

Thus only $V_{12} = \langle 210 | H_1 | 200 \rangle = \langle 200 | H_1 | 210 \rangle$ are finite and the secular equation that determines the perturbative correction is given by

$$\begin{vmatrix} E_2^0 - E & V_{12} & 0 & 0 \\ V_{12} & E_2^0 - E & 0 & 0 \\ 0 & 0 & E_2^0 - E & 0 \\ 0 & 0 & 0 & E_2^0 - E \end{vmatrix} = 0, \tag{9.90}$$

where E_2^0 is the energy of the unperturbed $n = 2$ hydrogen level. The four basis states are arranged in the order $(|200\rangle, |210\rangle, |211\rangle, |21\bar{1}\rangle)$.

By solving the secular equation, we find the perturbed energy levels

$$E_a = E_2^0 + V_{12}, \tag{9.91}$$
$$E_b = E_2^0 - V_{12}, \tag{9.92}$$
$$E_c = E_d = E_2^0. \tag{9.93}$$

The values of the energy for the states, $|211\rangle$ and $|21-1\rangle$, are unchanged. This is because the electric field does not break all the symmetries. It just reduces the spherical symmetry to a cylindrical symmetry. The corrections to the energy levels of the other two states, $|200\rangle$ and $|210\rangle$, are given by $\pm V_{12}$.

The linear level splitting by an external electric field appears only in hydrogen. It results from the linearity of V_{12} in the electric field E.

In a many-electron system, the l degeneracy is lifted. Therefore, on average the dipole moment vanishes and the first order perturbation becomes zero. But the atom is polarized by the external field. Such field-induced dipole moment is proportional to the field strength, thus the correction results from the second order perturbation, and the energy of the atom changes with E^2. This phenomenon is called the quadratic Stark effect.

9.5 PROBLEMS

1. Consider a particle moving in a one-dimensional potential $V(x)$ which is less than zero in certain ranges of space, i.e. $V(x) \leq 0$, and $V(x \to \pm\infty) \to 0$. Show that there is at least one bound state.

   Hint: There are more than one methods to prove this statement. One of them is to use the variational principle with a normalizable positive trial wave function, such as

   $$\Psi(x) = e^{-ax^2}.$$

2. A particle moving in a one-dimensional potential of harmonic oscillator

   $$V(x) = \frac{1}{2}\mu\omega^2 x^2.$$

   Using the trial wave function

   $$\Psi(x) = \begin{cases} C\left(1 - \dfrac{x^2}{a^2}\right), & |x| < a, \\ 0, & |x| \geq a \end{cases}$$

   to estimate the ground state energy. $a > 0$ is the variational parameter and C is the normalization constant.

3. The radial wave function of electron in Hydrogen atom, $R(r)$, satisfies the equation

   $$\left[-\frac{\hbar^2}{2\mu r}\frac{d^2}{dr^2}r + \frac{l(l+1)\hbar^2}{2\mu r^2} - \frac{e^2}{4\pi\varepsilon_0 r} \right] R(r) = ER(r).$$

   Using the trial wave function

   $$R(r) = Cr^l e^{-\lambda r}$$

   to estimate the lowest energies of the $l = 0$ and $l = 1$ states, and compare with the exact results.

4. Use the variational method to estimate the ground state energy of a particle in the triangular potential well

   $$V(x) = \begin{cases} V_0 x, & x > 0, \\ +\infty, & x < 0. \end{cases}$$

   (Hint: In the construction of the trial wave function, consider the behavior of the wave function in the limit $x \to \infty$ and at $x = 0$.)

5. Nuclear force: A deuteron is a bound state of a neutron and a proton, which interact through a strong short-range potential. Reasonable approximates to this potential are
   (1) Yukawa potential

   $$V(r) = -\frac{V_0 R}{r}e^{-r/R}.$$

(2) Fermi potential

$$V(r) = \begin{cases} -V_0, & r < R, \\ 0 & r > R. \end{cases}$$

a. Use $\Psi(\mathbf{r}) = \exp(-r/2R)$ as a trial function to find an expression for the ground state energy E_0 in each case.

b. Taking $E_0 = 3.5 \times 10^{-13}$ Joules and $R = 2 \times 10^{-15}$m, find V_0 in each case.

6. A particle in a two-dimensional infinite potential well

$$V(x,y) = \begin{cases} 0, & 0 < x < a \text{ and } 0 < y < a, \\ +\infty & \text{otherwise} \end{cases}$$

a. Find the eigenenergies and the corresponding eigenvectors for the three lowest energy levels?

b. Add a perturbation to the above potential

$$V_1(x,y) = \begin{cases} V_0 xy, & 0 < x < a \text{ and } 0 < y < a, \\ 0 & \text{otherwise} \end{cases}$$

Calculate the first order energy corrections to the first lowest energy levels caused by this perturbation.

7. Consider a quantum system with just three linear independent states, the unperturbed and perturbation Hamiltonians are defined respectively by

$$H_0 = \begin{pmatrix} 0 & 1 & 0 \\ 1 & 0 & 0 \\ 0 & 0 & -1 \end{pmatrix}, \qquad H_1 = \alpha \begin{pmatrix} 1 & 0 & 0 \\ 0 & 0 & 1 \\ 0 & 1 & 0 \end{pmatrix}$$

where α is a small number.

a. Write down all the eigenstates and the corresponding eigenvalues of the unperturbed Hamiltonian.

b. Solve for the exact eigenvalues of the full Hamiltonian, $H = H_0 + H_1$.

c. For the nondegenerate eigenstates of H_0, find the approximate eigenvalues up to the second order approximation of α.

d. Use degenerate perturbation theory to find the first order correction to the degenerate eigenstates (if exist) of H_0.

8. A spin-1 ion located in a crystal with rhombic symmetry is effectively described by the Hamiltonian

$$H = \alpha S_z^2 + \beta \left(S_x^2 - S_y^2 \right), \qquad (\beta \ll \alpha),$$

where (S_x, S_y, S_z) are the three components of the $S = 1$ spin operator. Take the α-term as the unperturbed Hamiltonian, and the β-term as the perturbation.

a. Find the eigenvalues and eigenvectors of the unperturbed Hamiltonian.

  b. Find the perturbed energy levels to first order.
9. A charged particle is in one of the three first excitation states of the three-dimensional Harmonic oscillator described by the Hamiltonian

$$H = -\frac{\hbar^2}{2\mu}\nabla^2 + \frac{1}{2}\mu\omega^2 r^2.$$

Now turn on a weak magnetic field along the z-axis which applies a Zeeman interaction to the particle

$$H_1 = -\alpha L_z = -\alpha(xp_y - yp_x),$$

where α is a coupling constant proportional to the applied magnetic field. Find the level splitting caused by H_1 to the leading order in α.

10. Each eigenfunction of the unperturbed Hamiltonian H_0 is also a parity eigenfunction, show that the eigenfunctions of the Hamiltonian $H = H_0 + H_1$ that are correct to first order in perturbation theory are not parity eigenfunctions unless H_1 commutes with the parity operator.

11. The hyperfine interaction between the electron at a s-orbital and the proton in a hydrogen atom is described by the Hamiltonian

$$H = A\sigma_e \cdot \sigma_p \delta(\mathbf{r})$$

where A is the coupling constant, σ_e and σ_p are the Pauli matrices of the electron and proton, respectively. The orbital state of the electron is assumed to be at the $|nlm\rangle = |100\rangle$ state. Calculate the hyperfine splitting to this energy state caused by the above interaction to the leading order approximation.

10 Quantum transition

10.1 TIME-DEPENDENT SCHRÖDINGER EQUATION

Let us consider a system in which the perturbation is time-dependent

$$H = H_0 + \lambda H_1(t). \tag{10.1}$$

Again a small parameter λ is used to identify the orders of perturbation. It is set to 1 in the final expansion. The unperturbed Hamiltonian H_0 is assumed time-independent and can be diagonalized. The eigenvalues and the corresponding orthonormal eigenfunctions of H_0 are given by E_n and $|n\rangle$, respectively

$$H_0 |n\rangle = E_n |n\rangle. \tag{10.2}$$

Since all the eigenfunctions $|n\rangle$ form a complete basis set, the general solution of the Schrödinger equation

$$i\hbar \frac{\partial}{\partial t} |\Psi(t)\rangle = H |\Psi(t)\rangle \tag{10.3}$$

can be expanded as

$$|\Psi(t)\rangle = \sum_n C_n(t) e^{-iE_n t/\hbar} |n\rangle, \tag{10.4}$$

where C_n are time-dependent coefficients. If $|\Psi(t)\rangle$ is normalized, then

$$\sum_n |C_n(t)|^2 = 1 \tag{10.5}$$

and

$$C_n(t) = \langle n|\Psi(t)\rangle \tag{10.6}$$

is the amplitude of finding the system in the unperturbed state $|n\rangle$ at time t.

Substituting Eq. (10.4) into Eq. (10.3), we obtain the equation that determines the coefficients

$$\boxed{\frac{\partial C_n(t)}{\partial t} = \frac{\lambda}{i\hbar} \sum_m C_m(t) e^{i\omega_{nm}t} \langle n| H_1(t) |m\rangle,} \tag{10.7}$$

where

$$\boxed{\omega_{nm} = \frac{E_n - E_m}{\hbar}.} \tag{10.8}$$

is the Bohr angular frequency.

Eq. (10.7) is strictly equivalent to the Schrödinger equation. There is no approximation used in obtaining equation. It can be used to solve rigorously, for example, the evolution of quantum states in a two-level system with time-dependent interactions. However, it is generally difficult to solve this equation rigorously.

DOI: 10.1201/9781003174882-10

10.1.1 PERTURBATION EXPANSION

Now we solve Eq. (10.7) perturbatively. We expand the coefficients $C_n(t)$ in powers of λ as

$$C_n(t) = C_n^0(t) + \lambda C_n^1(t) + \cdots. \tag{10.9}$$

It should be noted that the superscript i in C_n^i is not an exponent. It just denotes the order of expansion. Substituting the above expression into Eq. (10.7) and equating the coefficients of equal powers of λ, we find that up to the order of λ^2

$$\frac{d}{dt}C_n^0(t) = 0, \tag{10.10}$$

$$\frac{d}{dt}C_n^1(t) = \frac{1}{i\hbar}\sum_m C_m^0(t)\,e^{i\omega_{nm}t}\,\langle n|H_1(t)|m\rangle, \tag{10.11}$$

$$\frac{d}{dt}C_n^2(t) = \frac{1}{i\hbar}\sum_m C_m^1(t)\,e^{i\omega_{nm}t}\,\langle n|H_1(t)|m\rangle. \tag{10.12}$$

The first equation indicates that the zeroth order of coefficients $C_n^{(0)}$ are time-independent. In general, the jth-order correction is determined by

$$\frac{d}{dt}C_n^j(t) = \frac{1}{i\hbar}\sum_m C_m^{j-1}(t)\,e^{i\omega_{nm}t}\,\langle n|H_1(t)|m\rangle. \tag{10.13}$$

10.1.2 FIRST ORDER CORRECTION

In Eq. (10.10), if the initial state at $t=0$ is in a particular unperturbed state $|k\rangle$ of energy E_k, we have

$$C_n^0(t) = \delta_{n,k}. \tag{10.14}$$

Eq. (10.11) then becomes

$$\frac{d}{dt}C_n^1(t) = \frac{1}{i\hbar}e^{i\omega_{nk}t}\,\langle n|H_1(t)|k\rangle. \tag{10.15}$$

The solution of this equation is given by

$$C_n^1(t) = \frac{1}{i\hbar}\int_0^t dt\,e^{i\omega_{nk}t}\,\langle n|H_1(t)|k\rangle. \tag{10.16}$$

To first order in perturbation, the transition probability corresponding to the transition from $|k\rangle$ to $|n\rangle$, namely the probability that the system, initially in the state k, being found at time t in the state $n \neq k$, is given by

$$P_{nk}(t) = |C_n^1(t)|^2 = \frac{1}{\hbar^2}\left|\int_0^t dt\,e^{i\omega_{nk}t}\,\langle n|H_1(t)|k\rangle\right|^2, \qquad n \neq k. \tag{10.17}$$

This equation indicates that the transition from $|k\rangle$ to $|n\rangle$ is forbidden if

$$\langle n|H_1(t)|k\rangle = 0, \tag{10.18}$$

which defines the selection rule between the quantum numbers of the initial and final states.

For the state k, to the first order approximation, the coefficient $C_k(t)$ is given by

$$
\begin{aligned}
C_k(t) &\approx C_k^0(t) + C_k^1(t) \\
&= 1 + \frac{1}{i\hbar} \int_0^t dt\, \langle k|H_1(t)|k\rangle \\
&\approx \exp\left[\frac{1}{i\hbar} \int_0^t dt\, \langle k|H_1(t)|k\rangle\right]
\end{aligned}
\tag{10.19}
$$

so that

$$|C_k(t)|^2 \approx 1. \tag{10.20}$$

Thus the principal effect of the perturbation on the initial state is to change its phase.

10.1.3 TWO-LEVEL SYSTEMS

As an example, let us consider a two-level system, $|1\rangle$ and $|2\rangle$. They are the eigenstates of the unperturbed Hamiltonian H_0, with the corresponding eigenenergies E_1 and E_2 ($E_1 \neq E_2$)

$$
\begin{aligned}
H_0|1\rangle &= E_1|1\rangle, & (10.21) \\
H_0|2\rangle &= E_2|2\rangle. & (10.22)
\end{aligned}
$$

At $t = 0$, the system is assumed to be at the first eigenstate, i.e.

$$|\Psi(t=0)\rangle = |1\rangle. \tag{10.23}$$

Now we turn on a time-dependent perturbation, $H_1(t)$, and assume that the diagonal matrix elements of $H_1(t)$ are zero

$$\langle 1|H_1(t)|1\rangle = \langle 2|H_1(t)|2\rangle = 0, \tag{10.24}$$

and that the offdiagonal matrix elements are nonzero

$$h_{12} = h_{21}^* = \langle 1|H_1(t)|2\rangle. \tag{10.25}$$

For this two-level system, the wave function can be generally expressed as

$$|\Psi(t)\rangle = C_1(t)e^{-iE_1t/\hbar}|1\rangle + C_2(t)e^{-iE_2t/\hbar}|2\rangle. \tag{10.26}$$

Both $C_1(t)$ and $C_2(t)$ are time dependent. We determine them by perturbation expansion.

1. Zeroth order:
   Since the wave function is at the first eigenstate at $t = 0$, we have

$$C_1^0(t) = 1, \tag{10.27}$$
$$C_2^0(t) = 0. \tag{10.28}$$

2. First order correction:
   The first order corrections of the coefficients are determined by

$$\frac{\partial C_1^1(t)}{\partial t} = \frac{1}{i\hbar} C_2^0(t) e^{-i\omega t} h_{12}(t) = 0, \tag{10.29}$$

$$\frac{\partial C_2^1(t)}{\partial t} = \frac{1}{i\hbar} C_1^0(t) e^{i\omega t} h_{21}(t) = \frac{1}{i\hbar} e^{i\omega t} h_{21}(t), \tag{10.30}$$

where

$$\omega = \frac{E_2 - E_1}{\hbar} \tag{10.31}$$

is the angular frequency.
The solution of the above equations is

$$C_1^1(t) = 0, \tag{10.32}$$

$$C_2^1(t) = \frac{1}{i\hbar} \int_0^t dt' e^{i\omega t'} h_{21}(t') = \frac{1}{i\hbar} f(\omega, t), \tag{10.33}$$

where

$$f(\omega, t) = \int_0^t dt' e^{i\omega t'} h_{21}(t'). \tag{10.34}$$

3. Second order correction:
   The second order correction to the coefficient of the first state is determined by the equation

$$\frac{d}{dt} C_1^2(t) = \frac{1}{i\hbar} C_2^1(t) e^{-i\omega t} h_{12}(t)$$

$$= -\frac{1}{\hbar^2} e^{-i\omega t} h_{12}(t) f(\omega, t). \tag{10.35}$$

The solution is

$$C_1^2(t) = -\frac{1}{\hbar^2} \int_0^t dt' h_{12}(t') e^{-i\omega t'} f(\omega, t'). \tag{10.36}$$

The equation that determines the second order correction to the second state is

$$\frac{d}{dt} C_2^2(t) = \frac{1}{i\hbar} C_1^1(t) e^{i\omega t} h_{21}(t) = 0. \tag{10.37}$$

It yields

$$C_2^2(t) = 0. \tag{10.38}$$

Thus up to the second order perturbation, the coefficients are

$$C_1(t) \approx 1 - \frac{1}{\hbar^2} \int_0^t dt' h_{12}(t') e^{-i\omega t'} f(\omega, t'),$$ (10.39)

$$C_2(t) \approx \frac{1}{i\hbar} f(\omega, t).$$ (10.40)

This is an approximate solution, which violates the normalization condition

$$|C_1(t)|^2 + |C_2(t)|^2 \neq 1.$$ (10.41)

However, in principle, the normalization condition should be recovered if one can sum over all the orders of corrections.

10.2 MONOCHROMATIC PERTURBATION

In the case the perturbation is a monochromatic electromagnetic wave

$$H_1(t) = \begin{cases} H' \cos \omega_0 t, & t > 0, \\ 0, & t < 0, \end{cases}$$ (10.42)

with H' a time-independent constant, Eq. (10.16) becomes

$$\begin{aligned} C_n^1(t) &= \frac{1}{i\hbar} \langle n | H' | k \rangle \int_0^t dt\, e^{i\omega_{nk} t} \cos \omega_0 t \\ &= -\frac{1}{2\hbar} \langle n | H' | k \rangle \left(\frac{e^{i\omega_{nk}^+ t} - 1}{\omega_{nk}^+} + \frac{e^{i\omega_{nk}^- t} - 1}{\omega_{nk}^-} \right), \end{aligned}$$ (10.43)

where

$$\omega_{nk}^{\pm} = \omega_{nk} \pm \omega_0.$$ (10.44)

The transition probability from $|k\rangle$ to $|n\rangle$ $(n \neq k)$ is

$$P_{nk}(t) = \frac{1}{4\hbar^2} |\langle n | H' | k \rangle|^2 \left| \frac{e^{i\omega_{nk}^+ t} - 1}{\omega_{nk}^+} + \frac{e^{i\omega_{nk}^- t} - 1}{\omega_{nk}^-} \right|^2$$ (10.45)

The corresponding transition rate defined by

$$w_{nk}(t) = \frac{d}{dt} P_{nk}(t)$$ (10.46)

is

$$w_{nk}(t) = \frac{|\langle n | H' | k \rangle|^2}{2\hbar^2} \left[\sum_{\sigma=\pm} \left(\frac{\sin \omega_{nk}^{\sigma} t}{\omega_{nk}^{\sigma}} + \frac{\sin \omega_{nk}^{\sigma} t}{\omega_{nk}^{-\sigma}} \right) - \frac{2\omega_0 \sin 2\omega_0 t}{\omega_{nk}^+ \omega_{nk}^-} \right].$$ (10.47)

In the long time limit $\omega t \gg 2\pi$, $\sin \omega t$ is a fast oscillation function of time and the average of $\sin \omega t$ within a time interval $2\pi n/\omega$ and $2\pi (n+1)/\omega$ with n a large integer is zero except in the limit $\omega \to 0$.

Using the identity of the delta function

$$\delta(\omega) = \frac{1}{2\pi} \int_{-\infty}^{\infty} dp e^{ip\omega} = \left. \frac{\sin t\omega}{\pi\omega} \right|_{t\to\infty}, \tag{10.48}$$

we find the long-time transition rate to be

$$\boxed{w_{nk}(t) = \frac{\pi}{2\hbar^2} |\langle n| H' |k\rangle|^2 \left[\delta\left(\omega_{nk}^+\right) + \delta\left(\omega_{nk}^-\right) \right],} \tag{10.49}$$

which is time-independent. It indicates that in the long time limit, the transition happens in a uniform speed.

10.2.1 INTERACTION OF ATOMS WITH ELECTROMAGNETIC WAVE

The interaction of an electron with an applied electromagnetic field is governed by the gauge-invariant Hamiltonian,

$$H = \frac{1}{2\mu} (\mathbf{p} - e\mathbf{A})^2 = \frac{\mathbf{p}^2}{2\mu} - \frac{e}{2\mu} (\mathbf{p}\cdot\mathbf{A} + \mathbf{A}\cdot\mathbf{p}) + \frac{e^2}{2\mu} \mathbf{A}^2. \tag{10.50}$$

If we take the kinetic energy of the free particle, i.e. the first term on the right-hand side of the above equation, as the unperturbed Hamiltonian, the perturbation is simply given by

$$H_1 = -\frac{e}{2\mu} (\mathbf{p}\cdot\mathbf{A} + \mathbf{A}\cdot\mathbf{p}). \tag{10.51}$$

The $\mathbf{A}^2$-term does not interact with the particle and is omitted.

We assume $\mathbf{A}$ to be the gauge field generated by a monochromatic field of plane wave

$$\mathbf{A} = A_0 \hat{\varepsilon} \cos(\mathbf{k}\cdot\mathbf{r} - \omega_0 t). \tag{10.52}$$

A_0 is the amplitude and $\hat{\varepsilon}$ is a unit polarization vector. The corresponding electric field is

$$\mathbf{E} = -\frac{\partial \mathbf{A}}{\partial t} = A_0 \omega_0 \hat{\varepsilon} \sin(\mathbf{k}\cdot\mathbf{r} - \omega_0 t). \tag{10.53}$$

As the electromagnetic wave is a transverse wave, $\hat{\varepsilon}$ is perpendicular to the propagation direction, i.e. $\hat{\varepsilon}\cdot\mathbf{k} = 0$, so $\nabla\cdot\mathbf{A} = 0$ and H_1 can be simplified as

$$H_1 = -\frac{e}{\mu}\mathbf{A}\cdot\mathbf{p} = -\frac{e}{\mu}A_0(\hat{\varepsilon}\cdot\mathbf{p})\cos(\mathbf{k}\cdot\mathbf{r} - \omega_0 t). \tag{10.54}$$

In a hydrogenic system, the energy of the radiation field, $\hbar\omega_0$, must be of the order of atomic level spacing,

$$\hbar\omega_0 \sim \frac{Ze^2}{4\pi\varepsilon_0 R_{\text{atom}}}. \tag{10.55}$$

The corresponding wave length is

$$\lambda = \frac{2\pi c}{\omega_0} = \frac{2\pi R_{\text{atom}}}{Z\alpha} \approx \frac{274\pi R_{\text{atom}}}{Z}, \tag{10.56}$$

where

$$\alpha = \frac{e^2}{4\pi\varepsilon_0 c\hbar} \approx \frac{1}{137} \qquad (10.57)$$

is the fine structure constant. For a light atom (small Z), the wave length is much larger than the atom radius R_{atom}

$$\lambda \gg R_{\text{atom}}. \qquad (10.58)$$

Therefore, we can ignore the $\mathbf{r}$-dependence of the electromagnetic wave and approximate H_1 by

$$H_1 \approx -\frac{e}{\mu c} A_0 (\hat{\varepsilon} \cdot \mathbf{p}) \cos(\omega_0 t). \qquad (10.59)$$

This is just the monochromatic perturbation defined in Eq. (10.42) with

$$H' = -\frac{e}{\mu} A_0 (\hat{\varepsilon} \cdot \mathbf{p}). \qquad (10.60)$$

The matrix element of H' is

$$\langle n | H' | k \rangle = -\frac{e}{\mu} A_0 \langle n | \hat{\varepsilon} \cdot \mathbf{p} | k \rangle. \qquad (10.61)$$

Using the equation

$$[\mathbf{r}, H_0] = \frac{i\hbar}{\mu} \mathbf{p}, \qquad (10.62)$$

we can express this matrix element as

$$\langle n | H' | k \rangle = \frac{e}{i\hbar} A_0 (E_n - E_k) \hat{\varepsilon} \cdot \langle n | \mathbf{r} | k \rangle = -ieA_0 \omega_{nk} \langle n | \hat{\varepsilon} \cdot \mathbf{r} | k \rangle. \qquad (10.63)$$

Since $e\mathbf{r}$ is the dipole operator of electron, one can regard H' as an effective interaction induced by the electric dipole interaction. Therefore, the corresponding transition rate in the long time limit, according to Eq. (10.49), is

$$\begin{aligned} w_{nk}(t) &= \frac{\pi e^2 A_0^2 \omega_{nk}^2}{2\hbar^2} |\langle n | \hat{\varepsilon} \cdot \mathbf{r} | k \rangle|^2 [\delta(\omega_{nk} - \omega_0) + \delta(\omega_{nk} + \omega_0)] \\ &= \frac{\pi e^2 A_0^2 \omega_0^2}{2\hbar^2} |\langle n | \hat{\varepsilon} \cdot \mathbf{r} | k \rangle|^2 [\delta(\omega_{nk} - \omega_0) + \delta(\omega_{nk} + \omega_0)]. \end{aligned} \qquad (10.64)$$

10.2.2 ABSORPTION AND STIMULATED EMISSION

If both the initial and final energy levels are discretized, the delta function in Eq. (10.49) indicates that the transition happens only when the energy difference between the two levels matches the frequency of the perturbation

$$E_n - E_k = \pm\hbar\omega_0. \qquad (10.65)$$

This is a consequence of energy conservation.

Depending on the relative positions of the two levels, two processes may happen:

1. Absorption:
   The initial state is in the lower energy level $|d\rangle$, i.e. $|k\rangle = |d\rangle$ and $E_k = E_d$, and the final state is in the upper energy level $|u\rangle$, i.e. $|n\rangle = |u\rangle$ and $E_n = E_u > E_d$, w_{nk} becomes

$$w_{ud}(t) = \frac{\pi}{2\hbar}\left|\langle u|H'|d\rangle\right|^2 \delta\left(E_u - E_d - \hbar\omega_0\right). \qquad (10.66)$$

This corresponds to the process that the initial state absorbs the radiation wave of the frequency $\omega_0 = (E_u - E_d)/\hbar$ and jumps to the final state.

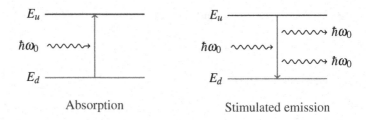

Absorption   Stimulated emission

2. Stimulated emission:
   The initial state is in the upper energy level, $|k\rangle = |u\rangle$ and $E_k = E_u$, and the final state is in the lower energy level $|n\rangle = |d\rangle$ and $E_n = E_d$. Now the transition rate w_{nk} is

$$w_{du}(t) = \frac{\pi}{2\hbar}\left|\langle d|H'|u\rangle\right|^2 \delta\left(E_u - E_d - \hbar\omega_0\right). \qquad (10.67)$$

This corresponds to the stimulated emission of the incident wave with the frequency $\omega_0 = (E_u - E_d)/\hbar$. It implies that, by shining a light wave on a particle, it can make a transition from a higher energy level E_k to a lower one E_n. This transition will emit another photon of the same frequency as the initial one to converse the energy. Thus there is one photon in, and two photons out. This raises the possibility of amplification of light. If there are many particles, all in the higher energy state E_k, which are triggered by a single photon with the proper frequency, a chain reaction would happen: the first photon produces two, these two photons produce four, and so on. In fact, this is the principle underlying the light amplification by stimulated emission of radiation (LASER). Of course, this coherent amplification can happen only the majority of atoms are in the higher energy state (this is also called population inversion). Otherwise, the absorption would reduce the number of photons and extinguish this amplification effect.

Comparing Eq. (10.67) with Eq. (10.66) and using the property $|\langle d|H'|u\rangle| = |\langle u|H'|d\rangle|$, we find that $w_{du} = w_{ud}$. Thus the transition rate of the stimulated emission equals that of the absorption, at least in the first order perturbation.

10.2.3 FERMI'S GOLDEN RULE

The delta function in Eq. (10.66) or (10.67) suggests that the absorption or stimulated emission happens only when the energy difference between the final and initial states matches precisely the frequency of the applied electromagnetic wave. Apparently, this is practically challenging if such a strict condition of transition should be satisfied. However, a discrete energy level is not mathematical line. Instead, it is always broadened by quantum fluctuations, collision with other atoms, and many other effects. Thus an energy level is quasicontinuum.

In general, if the final states fall into a continuum, such as in a continuous spectrum or a broadened energy level, the total transition rate from the initial state to the final state is given by

$$W_{k\to n}(t) = \int_{-\infty}^{\infty} dE \rho(E) w_{kn}(t) = \frac{\pi}{2\hbar^2} |\langle n| H'|k\rangle|^2 \rho(E_n), \qquad (10.68)$$

where $E_n = E_k - \hbar\omega_0$ or $E_n = E_k + \hbar\omega_0$. $\rho(E)$ is the density of energy levels at energy E and $\rho(E) dE$ is the number of energy states in an energy interval between $E - dE/2$ and $E + dE/2$. This formula was first derived by P. A. M. Dirac. It was later dubbed by E. Fermi "the golden rule".

The Fermi's golden rule removes the singularities in Eqs. (10.66) and (10.67). It describes the relationship between the transition rate, which is defined as the probability of a transition per unit time, from one energy eigenstate of a quantum system to a group of energy eigenstates in a continuum, as a result of a weak perturbation. This transition rate is proportional to the square of the matrix element of the perturbation as well as the density of states of the final states. Under a monochromatic perturbation, this rate is independent of time in the long-time limit.

The Fermi's golden rule can be also generalized to include the case that the frequency of the applied electromagnetic wave has a finite distribution.

10.2.4 SELECTION RULES*

The quantum transition induced by a monochromatic perturbation is determined by the matrix element of H' between the initial and final states, $\langle n|H'|k\rangle$. This matrix element is often zero in an atomic system with high symmetry. If we know in advance the rules between the quantum numbers of the initial and final states at which $\langle n|H'|k\rangle$ becomes zero, we would immediately know which transition is forbidden or allowed without evaluating explicitly the matrix element. These transition rules between the initial and final states are called selection rules.

Suppose we are interested in systems like hydrogen with spherical potential so that both the initial and final states are described by the hydrogenic wave function $|n, l, m\rangle$ with n the principal quantum number and (l, m) the two magnetic quantum numbers. In the dipole approximation, H' is proportional to $\hat{\varepsilon} \cdot \mathbf{r}$, we thus have

$$\langle n'l'm'|H'|nlm\rangle \propto \langle n'l'm'|\hat{\varepsilon} \cdot \mathbf{r}|nlm\rangle. \qquad (10.69)$$

Since $\mathbf{r} = (x - iy, z, x + iy)$ is a rank-1 spherical tensor, from the Wigner-Eckart theorem introduced in Section 5.6, we know that $\langle n'l'm'|\hat{\boldsymbol{\varepsilon}} \cdot \mathbf{r}|nlm\rangle \neq 0$ if the quantum numbers satisfy the following conditions

1. $l' - l = 0, \pm 1$.
2. $m' = m$, if the polarization is along the z-axis so that $\langle n'l'm'|\hat{\boldsymbol{\varepsilon}} \cdot \mathbf{r}|nlm\rangle = \langle n'l'm'|z|nlm\rangle$.
3. $m' = m \pm 1$, if the polarization of the light is along the x- or y-axis.

Furthermore, since the spherical harmonics $Y_{lm}(\theta, \phi)$, defined by Eq. (5.69) is odd or even under the transformation of spatial reflection $\mathbf{r} \rightarrow -\mathbf{r}$, or $(\theta, \phi) \rightarrow (\pi - \theta, \pi + \phi)$, depending on the value l

$$Y_{lm}(\pi - \theta, \pi + \phi) = (-)^l Y_{lm}(\theta, \phi), \tag{10.70}$$

and $\mathbf{r}$ is odd under the spatial reflection, the matrix element with $l' = l$ should vanish

$$\langle n'lm'|\hat{\boldsymbol{\varepsilon}} \cdot \mathbf{r}|nlm\rangle = 0. \tag{10.71}$$

Thus the selection rules for the electromagnetic transitions induced by the electric dipole interaction in hydrogenic atoms are

$$\Delta_l = l' - l = \pm 1, \tag{10.72}$$

and

$$\Delta_m = m' - m = \begin{cases} 0 & \text{if } \hat{\boldsymbol{\varepsilon}} = \hat{z}, \\ \pm 1 & \text{if } \hat{\boldsymbol{\varepsilon}} \perp \hat{z}. \end{cases} \tag{10.73}$$

10.2.5   CONSTANT PERTURBATION

In the limit $\omega_0 = 0$, the perturbation becomes

$$H_1(t) = \begin{cases} H', & t > 0, \\ 0, & t < 0. \end{cases} \tag{10.74}$$

It is a time-independent perturbation that is switched on at $t = 0$.

Again, from an initial eigenstate $|k\rangle$ of the unperturbed Hamiltonian, the first order correction to the wave function is

$$C_k^1(t) = \frac{t}{i\hbar} \langle k|H'|k\rangle, \tag{10.75}$$

$$C_n^1(t) = \frac{1 - e^{i\omega_{nk}t}}{\hbar \omega_{nk}} \langle n|H'|k\rangle, \quad n \neq k. \tag{10.76}$$

From Eq. (10.19), the coefficient C_k at the initial state to first order perturbation is found to be

$$C_k(t) \approx \exp\left[\frac{t}{i\hbar} \langle k|H'|k\rangle\right]. \tag{10.77}$$

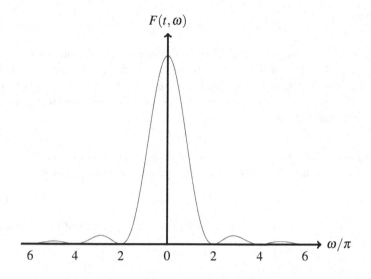

Figure 10.1 ω dependence of $F(t,\omega) = 2\sin^2(\omega t/2)/\omega^2$ for $t = 1$.

Hence the amplitude of the wave function at $|k\rangle$ is approximately given by

$$C_k(t)\,e^{-itE_k/\hbar} \approx e^{-iE_k't/\hbar}, \tag{10.78}$$

where

$$E_k' = E_k + \langle k|\,H'\,|k\rangle \tag{10.79}$$

is just the correction to the energy up to the first order perturbation, same as that obtained by the stationary perturbation theory.

First-order transition probability from $|k\rangle$ to $|n \neq k\rangle$ is

$$P_{nk}(t) = \frac{2}{\hbar^2}\,|\langle n|\,H'\,|k\rangle|^2\,F(t,\omega_{nk}), \tag{10.80}$$

where

$$F(t,\omega) = \frac{1-\cos(\omega t)}{\omega^2} = \frac{2\sin^2(\omega t/2)}{\omega^2}. \tag{10.81}$$

Figure 10.1 shows how $F(t,\omega)$ varies with ω. Given t, $F(t,\omega)$ oscillates with ω and shows a sharp peak at $\omega = 0$

$$F(t,\omega=0) = \frac{t^2}{2}. \tag{10.82}$$

The width is about $2\pi/t$.

To integrate $F(t,\omega)$ with respect to ω, we obtain

$$\int_{-\infty}^{\infty} d\omega\,F(t,\omega) = \pi t. \tag{10.83}$$

In the limit $t \to \infty$,

$$F(t, \omega) \sim \pi t \delta(\omega). \tag{10.84}$$

Since $F(t, \omega_{nk})$ is sharply peaked about the value $\omega_{nk} = 0$, with a width approximately given by $2\pi/t$, it is clear that transitions to those final state $|n\rangle$ for which ω_{nk} does not deviate from zero by more than $\delta\omega_{nk} \approx 2\pi/t$ will be strongly favoured. Hence the transitions $k \to n$ occur mainly towards the final state whose energy is located in a band of width

$$\Delta E \approx 2\pi\hbar/t \tag{10.85}$$

about the initial energy E_k, so that the unperturbed energy is conserved within $2\pi\hbar/t$. This result can be understood from the time-energy uncertainty principle

$$\Delta E \Delta t > \hbar. \tag{10.86}$$

Now let us examine the transition probability as a function of t, and distinguish two cases:

1. Case I: $\omega_{nk} \neq 0$

   If the state n is not degenerate with the initial state k, the transition probability is given by Eq. (10.80), which oscillates with a period $T = 2\pi/|\omega_{nk}|$. For small t with respect to the period of oscillation, the transition probability is given by

   $$P_{nk}(t) \approx \frac{t^2}{\hbar^2}|\langle n|H'|k\rangle|^2, \tag{10.87}$$

   which increases quadratically with time.
   The total transition probability from the initial state to all other states is given by

   $$P(t) = \sum_{n \neq k} \frac{2}{\hbar^2}|\langle n|H'|k\rangle|^2 \frac{2\sin^2(\omega_{nk}t/2)}{\omega_{nk}^2}. \tag{10.88}$$

   Clearly, in order that the perturbation is valid, it is necessary that $P(t) \ll 1$. A sufficient condition for the validity of inequality is

   $$\sum_{n \neq k} \frac{4}{(E_n - E_k)^2}|\langle n|H'|k\rangle|^2 \ll 1. \tag{10.89}$$

   This inequality is satisfied if the perturbation H' is sufficiently weak.

2. Case II: $\omega_{nk} = 0$

   If the state n is degenerate with the initial state, $F(t, \omega_{nk}) = t^2/2$, we have

   $$P_{nk}(t) = \frac{t^2}{\hbar^2}|\langle n|H'|k\rangle|^2. \tag{10.90}$$

   $P_{nk}(t)$ increases indefinitely with time. Thus after a sufficient length of time the transition probability will eventually exceed unity, which is clearly absurd. We thus conclude that the present perturbation treatment cannot be applied to degenerate systems over long periods of time.

10.3 EINSTEIN'S THEORY OF RADIATION

Besides the absorption and stimulated emission of photons discussed in Section 10.2.2, an electron can also transit from an excited energy state to a lower energy state, for example the ground state, and emits a photon with a quantized amount of energy equal to the difference between the two energy levels even without an external trigger. This is just the spontaneous emission of photons. In fact, the spontaneous emission is ubiquitous and responsible for most of the light we see around us.

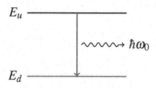

Spontaneous emission

A spontaneous emission of photon is a probabilistic event, resulting from the electromagnetic fluctuation of vacuum, because the "vacuum" of atoms or molecules is not truly empty. It is natural to interpret the spontaneous emission as a "stimulated emission" of photons driven by the zero point oscillation of electromagnetic field of in the empty space, which is generally referred to as vacuum fluctuation. However, unlike in the true stimulated emission where the emitted photons keep the phases as well as the propagation directions as the incident ones, the phase of the photon in spontaneous emission is random as is the direction in which the photon propagates. So the spontaneous emission does not have the amplification effect.

In 1917, Einstein analyzed the processes of radiations in the formation of atomic spectra, before the theory of quantum mechanics was established. He did not identify the microscopic mechanics responsible for the three processes. Nevertheless, he introduced for the first time the process of stimulated emission and predicted that it coexists with the spontaneous emission in order to reproduce Planck's formula of radiation.

To elaborate the picture of Einstein, let us consider a system of atoms, N_u of them in the upper energy state $|u\rangle$ and N_d of them in the lower energy state $|d\rangle$. Assuming A to be the spontaneous emission rate, so that the number of atoms spontaneously leaving the higher energy state to the lower energy state per unit time is $N_u A$. The stimulated emission or absorption is proportional to the transition rate w_{ud} or w_{du}, as well as the energy intensity $I(\omega)$ $(\omega = (E_u - E_d)/\hbar)$ of the radiation waves.

Suppose these atoms with the radiation field are in thermodynamic equilibrium, so that the number of atoms in each energy level does not change with time, then the number of emissions should equal the number of absorption

$$AN_u + w_{ud}I(\omega)N_u = w_{du}I(\omega)N_d. \tag{10.91}$$

The spontaneous emission is an intrinsic property of atoms, and its decay rate is assumed to be independent of the intensity of radiation field. From the above equation, the intensity of the radiation field is found to be

$$I(\omega) = \frac{AN_u}{w_{du}N_d - w_{ud}N_u}.$$ (10.92)

Since the distribution of the atoms, in thermodynamic equilibrium at temperature T, is described by the Boltzmann statistics, so

$$\frac{N_d}{N_u} = \exp\left(-\frac{E_d - E_u}{k_B T}\right) = \exp\left(\frac{\hbar\omega}{k_B T}\right),$$ (10.93)

where k_B is the Boltzmann constant. Eq. (10.92) then becomes

$$I(\omega) = \frac{(A/w_{du})}{e^{\hbar\omega/k_B T} - (w_{ud}/w_{du})}.$$ (10.94)

By comparing it with the Planck's law of black body radiation,

$$I(\omega) = \frac{8\pi h}{c^3} \frac{\nu^3}{e^{h\nu/k_B T} - 1} = \frac{2\hbar}{\pi c^3} \frac{\omega^3}{e^{\hbar\omega/k_B T} - 1}.$$ (10.95)

we obtain two results that were first predicted by Einstein

1. The transition rate for stimulated emission is the same as for absorption.

$$w_{ud} = w_{du}.$$ (10.96)

   This agrees with the results, as discussed in Section 10.2.2, obtained from the first order perturbation calculation.
2. The spontaneous emission rate is related to the stimulated emission rate by the formula,

$$A = \frac{2\hbar\omega^3}{\pi c^3} w_{du}.$$ (10.97)

   It indicates that the spontaneous emission is intimately connected to the stimulated emission. This formula can be also used to determine the spontaneous emission rate A from the absorption or stimulated emission rate.

10.4 PROBLEMS

1. A two-level system is initially in the ground state of the Hamiltonian

$$H_0 = \begin{pmatrix} E_1 & 0 \\ 0 & E_2 \end{pmatrix}, \qquad (E_1 < E_2).$$

At time $t = 0$, a perturbation

$$H_1(t) = \begin{pmatrix} 0 & a \\ a & 0 \end{pmatrix} e^{-t/T}$$

is applied to the system. Find the probability that the system is in the excitation state in the long time limit, and discuss the condition for the validity of the first order perturbation.

2. Consider a particle of charge e and mass μ, which is in the ground state of one-dimensional harmonic oscillator described by the Hamiltonian

$$H_0 = \frac{p^2}{2\mu} + \frac{1}{2}\mu\omega^2 x^2.$$

A homogeneous electric field $\mathscr{E}(t)$ directed along the x-axis is switched on at time $t = 0$, so that the system is perturbed by the interaction

$$H_1(t) = -ex\mathscr{E}(t).$$

a. If $\mathscr{E}(t) = E_0$, a spatial and time independent constant, find the probability that the particle remains in the ground state after the field is switched on using perturbation theory, and compare with the exact result.

b. On the other hand, if $\mathscr{E}(t)$ has the form

$$\mathscr{E}(t) = E_0 e^{-t/\tau},$$

Find the probability of the particle being excited to an excitation state in the limit $t \to \infty$.

3. A hydrogen atom is at the ground state. Suddenly a pulse electric field, along the z-axis, is added to the atom. The potential energy generated by this pulse field is

$$H_1 = e\mathbf{E} \cdot \mathbf{r}\delta(t) = eEz\delta(t),$$

where E is a constant. Using perturbation theory to find the probability that the electron remains at the ground state after the pulse field is applied to the system.

4. A particle is in the ground state of a one-dimensional infinite square well with walls at $x = \pm L/2$. At time $t = 0$, the width of the well is suddenly increased to $2L$. Find the probability that the particle will be found in the nth eigenstate of the expended well.

5. Magnetic resonance: A spin-1/2 particle with gyromagnetic ratio γ, at rest in a static magnetic field $B_0\hat{z}$. At time $t = 0$, a small transverse radio-frequency field is switched on, so that the total field is

$$\mathbf{B}(t) = B_r\cos(\omega t)\hat{x} - B_r\sin(\omega t)\hat{y} + B_0\hat{z},$$

and the interaction is given by

$$H(t) = -\frac{1}{2}\gamma\hbar\sigma \cdot \mathbf{B}(t),$$

where σ are the Pauli matrices.

a. Write down the 2×2 Hamiltonian matrix.
b. If

$$|\Psi(t=0)\rangle = \begin{pmatrix} a_0 \\ b_0 \end{pmatrix}$$

is the spin state at $t = 0$, find how the spin state changes with time.

11 Adiabatic and diabatic Evolution

11.1 ADIABATIC VERSUS SUDDEN APPROXIMATIONS

If the Hamiltonian is time-dependent, the solution of the time-dependent Schrödinger equation in general becomes difficult. Nevertheless, we can still diagonalize the Hamiltonian at each given time to find its instantaneous eigenvalues $E_n(t)$ and the corresponding eigenvectors $|n,t\rangle$

$$H(t)|n,t\rangle = E_n(t)|n,t\rangle. \tag{11.1}$$

Clearly, both the eigenvalues E_n and eigenvectors $|n\rangle$ are time-dependent. These eigenstates still constitute a complete orthonormal basis set so that the general solution of the time-dependent Schrödinger equation

$$i\hbar \frac{\partial}{\partial t}|\Psi(t)\rangle = H(t)|\Psi(t)\rangle \tag{11.2}$$

can be expanded as

$$|\Psi(t)\rangle = \sum_n C_n(t)e^{i\theta_n(t)}|n,t\rangle, \tag{11.3}$$

where

$$\boxed{\theta_n(t) = -\frac{1}{\hbar}\int_0^t E_n(t)dt} \tag{11.4}$$

is called the dynamic phase.

Coefficients $C_n(t)$ can be determined by substituting Eq. (11.3) into Eq. (11.2). This leads to

$$\sum_n e^{i\theta_n(t)}\left[|n,t\rangle \partial_t C_n(t) + C_n(t)\partial_t|n,t\rangle\right] = 0. \tag{11.5}$$

Taking the inner product with $\langle m,t|$, the above equation becomes

$$\partial_t C_m(t) = -\sum_n e^{i\theta_n(t)-i\theta_m(t)}C_n(t)\langle m,t|\partial_t|n,t\rangle. \tag{11.6}$$

By differentiating Eq. (11.1) with respect to time, we have

$$(\partial_t H)|n,t\rangle + H(t)\partial_t|n,t\rangle = (\partial_t E_n)|n,t\rangle + E_n(t)\partial_t|n,t\rangle, \tag{11.7}$$

hence

$$[E_n(t)-E_m(t)]\langle m,t|\partial_t|n,t\rangle = -\delta_{mn}\partial_t E_n + \langle m,t|(\partial_t H)|n,t\rangle, \tag{11.8}$$

DOI: 10.1201/9781003174882-11

and

$$\langle m,t| \partial_t |n,t\rangle = \frac{1}{E_n(t) - E_m(t)} \langle m,t| (\partial_t H) |n,t\rangle, \qquad (m \neq n). \qquad (11.9)$$

Substituting it into (11.6), we obtain

$$\partial_t C_m(t) = -C_m \langle m,t| \partial_t |m,t\rangle - \sum_{n \neq m} e^{i(\theta_n - \theta_m)} C_n(t) \frac{\langle m,t| (\partial_t H) |n,t\rangle}{E_n(t) - E_m(t)}. \qquad (11.10)$$

This is a rigourous result without involving any approximation.

11.1.1 THE ADIABATIC THEOREM

If the Hamiltonian $H(t)$ changes very slowly with time, so that the second term in Eq. (11.10) can be dropped out. In this case, Eq. (11.10) becomes

$$\partial_t C_m(t) \approx -C_m \langle m(t)| \partial_t |m(t)\rangle. \qquad (11.11)$$

Its solution is given by

$$C_m(t) \approx C_m(0) e^{i\gamma_m(t)}, \qquad (11.12)$$

where $\gamma_m(t)$ is called the geometric phase

$$\gamma_m(t) = i \int_0^t \langle m(t)| \partial_t |m(t)\rangle dt. \qquad (11.13)$$

In obtaining these results, we assume that all the eigenstates of $H(t)$ are discrete and nondegenerate.

Furthermore, if the particle is initially in the n'th instantaneous eigenstate at time $t = 0$, namely $C_m(0) = \delta_{mn}$, then

$$|\Psi(t)\rangle \approx e^{i\theta_n(t) + i\gamma_n(t)} |n(t)\rangle. \qquad (11.14)$$

So the time evolution of the state remains in the n'th instantaneous eigenstate of the time-dependent Hamiltonian, apart from two phase factors. This is just the adiabatic theorem first established by Max Born and Vladimir Fock in 1928, which states that

> Adiabatic theorem:
> A physical system with a time-dependent Hamiltonian $H(t)$ remains in its instantaneous eigenstate of $H(t)$, if $H(t)$ changes slowly enough with time and if there is a gap between the eigenvalue and the rest eigenvalues of $H(t)$.

The adiabatic theorem holds when the second term in Eq. (11.10) can be neglected. It is characterized by the probability that the final state of the system is different from the initial one, which determines the diabatic transition probability between different states. This transition is known as the Landau-Zener transition.

11.1.2 SUDDEN APPROXIMATION

On the contrary, if the Hamiltonian changes very quickly, there is no time for the system to adjust to the change. This leaves the system in the same state just before the change, which is called the sudden approximation.

Again, let us consider a Hamiltonian which changes continuously from an initial value H_i at time $t = 0$ to a final one H_f at $t = \tau$. The sudden approximation, which is a limit of the general diabatic approximation, is quantified by the probability of finding the system in a state other than that in which it started

$$\zeta = \langle \Psi(\tau) | \Psi(\tau) \rangle - \langle \Psi(\tau) | 0 \rangle \langle 0 | \Psi(\tau) \rangle, \tag{11.15}$$

where $|0\rangle = |\Psi(t=0)\rangle$ is the initial wave function.

The evolution of the wave function is determined by the integral equation

$$
\begin{aligned}
|\Psi(\tau)\rangle &= \exp\left(\int_0^\tau \frac{1}{i\hbar} H(t)\, dt\right) |0\rangle \\
&= \exp\left(\frac{\tau \overline{H}}{i\hbar}\right) |0\rangle \\
&= \left(1 + \frac{\tau \overline{H}}{i\hbar} - \frac{\tau^2 \overline{H}^2}{2\hbar^2} + \cdots\right) |0\rangle,
\end{aligned} \tag{11.16}
$$

where

$$\overline{H} = \frac{1}{\tau} \int_0^\tau H(t)\, dt \tag{11.17}$$

is the average Hamiltonian over the time interval $(0, \tau)$.

Substituting the above expression into Eq. (11.15) and keeping the terms up to the order of τ^2, then ζ is found to be

$$\boxed{\zeta = \frac{\tau^2}{\hbar^2}\left(\langle 0|\overline{H}^2|0\rangle - \langle 0|\overline{H}|0\rangle^2\right) = \frac{\tau^2}{\hbar^2}\left(\Delta\overline{H}\right)^2,} \tag{11.18}$$

$\Delta\overline{H}$ is the root mean square deviation of $\overline{H}$. It measures the fluctuation of energy in the transition period in the initial state.

The sudden approximation is valid when $\zeta \ll 1$, namely when the probability of finding the system not in a state it started approaches zero. The validity condition of the sudden approximation is quantified by the probability that the system remains unchanged, $1 - \zeta$, or by the condition

$$\tau \cdot \Delta\overline{H} \ll \hbar. \tag{11.19}$$

Under this approximation, the wave function at $t = \tau$ equals approximately the initial one

$$|\Psi(\tau)\rangle \approx |0\rangle. \tag{11.20}$$

11.1.3 QUANTUM ZENO EFFECT

Quantum zeno effect, named by Misra and Sudarshan, is an inhibition of induced or spontaneous transition between quantum states by frequent measurements, so that the system remains in its initial state throughout a long time interval. In other words, the frequent measurements decouple the interaction between the system and the environment, and suppress the evolution of quantum state. This effect is also called the Turing paradox.

To illustrate the key feature of quantum Zeno effect, let us consider a system that is initially in the n'th eigenstate, $|n\rangle$, of an observable O to which the measurement is performed. Each measurement is done at each time interval τ. Initially, the evolution of the state from $t = 0$ to $t = \tau$ is governed by the Schrödinger equation,

$$|\Psi(\tau)\rangle = \exp\left(-\frac{iH\tau}{\hbar}\right)|n\rangle. \tag{11.21}$$

The Hamiltonian is assumed to be time-independent. For sufficiently small τ, we may expend the above exponent into a power series,

$$|\Psi(\tau)\rangle \approx \left[1 - \frac{iH\tau}{\hbar} - \frac{H^2\tau^2}{2\hbar^2} + \cdots\right]|n\rangle. \tag{11.22}$$

After the first measurement, the probability of the system remaining in the initial state is

$$
\begin{aligned}
P_n(\tau) &= |\langle n|\Psi(\tau)\rangle|^2 \\
&= \left|1 - \frac{i\tau}{\hbar}\langle n|H|n\rangle - \frac{\tau^2}{2\hbar^2}\langle n|H^2|n\rangle + \cdots\right|^2 \\
&\approx \left|1 - \frac{\tau^2}{2\hbar^2}\langle n|H^2|n\rangle\right|^2 + \frac{\tau^2}{\hbar^2}\langle n|H|n\rangle^2 \\
&\approx 1 - \frac{\tau^2\Delta(H_n)^2}{\hbar^2},
\end{aligned}
\tag{11.23}
$$

where

$$\Delta H_n = \sqrt{\langle n|H^2|n\rangle - \langle n|H|n\rangle^2} \tag{11.24}$$

is the root mean square deviation of energy in the state $|n\rangle$. The probability of the system collapses to the other eigenstates of O is

$$1 - P_n(\tau) \approx \frac{\tau^2\Delta(H_n)^2}{\hbar^2}. \tag{11.25}$$

After an uninterrupted evolution from $t = \tau$ to $t = 2\tau$, the second measurement is taken. The starting state could be either in the original $|n\rangle$ state or in other eigenstates of O. After the second measurement, a lower bound of the probability of the system collapsed onto the state $|n\rangle$ is given by

$$P_n(2\tau) \geq P_n^2(\tau). \tag{11.26}$$

This is a lower bound because this probability does not include the contribution from the measurement to the states which are not in the n'th eigenstate of O after the first measurement at $t = \tau$.

We repeat the above measurement for N times and at each time allow the system to evolve without interruption for a time interval τ, the lower bound of the probability of the system remains in the initial state becomes

$$P_n(N\tau) \geq P_n^N(\tau) \approx \left[1 - \frac{\tau^2(\Delta H_n)^2}{\hbar^2}\right]^N. \tag{11.27}$$

In the limit $\tau \to 0$, $N \to \infty$ and total evolution time $t = N\tau$ is finite,

$$\lim_{\tau \to 0} P_n(t) \geq \left[1 - \frac{\tau^2(\Delta H_n)^2}{\hbar^2}\right]^{t/\tau} \approx \exp\left[-\frac{\tau t(\Delta H_n)^2}{\hbar^2}\right] \approx 1. \tag{11.28}$$

Apparently, the probability cannot be higher than 1. This inequality simply means that the frequent measurements will froze the system in its initial state and suppress completely the probability of transition from $|n\rangle$ to other eigenstates of O. It sounds that the frequent measurements plays a similar role as the sudden approximation. But physically they are completely different, because the sudden approximation does not involve the interaction with the measurement observers.

11.2 LANDAU-ZENER TRANSITION

11.2.1 RABI OSCILLATION

As an example of the adiabatic or diabatic evolution, let us consider a half-spin system subjected to the Zeeman interaction applied by two perpendicular magnetic fields.

We first add a transverse magnetic field B_x to the system. The Zeeman interaction induced by the applied field is described by the Hamiltonian

$$H_0 = \frac{1}{2}g\mu_B\sigma B_x = \hbar\omega\sigma_x = \begin{pmatrix} 0, & \hbar\omega \\ \hbar\omega, & 0 \end{pmatrix}, \tag{11.29}$$

where μ_B is the Bohr magneton, g is the Landé g-factor, and

$$\omega = \frac{g\mu_B B_x}{2\hbar} \tag{11.30}$$

is the angular frequency introduced to quantify the energy scale of the Zeeman interaction induced by the transverse field.

H_0 has two eigenvalues, $\pm\hbar\omega$, and the corresponding eigenvectors are

$$|-\rangle = \frac{1}{\sqrt{2}}(|\uparrow\rangle - |\downarrow\rangle), \qquad |+\rangle = \frac{1}{\sqrt{2}}|\uparrow\rangle + |\downarrow\rangle), \tag{11.31}$$

where $|\uparrow\rangle$ and $|\downarrow\rangle$ are the eigenstates of σ_z

$$|\uparrow\rangle = \begin{pmatrix} 1 \\ 0 \end{pmatrix}, \qquad |\downarrow\rangle = \begin{pmatrix} 0 \\ 1 \end{pmatrix}. \tag{11.32}$$

If the system is initially at a spin up state $|\uparrow\rangle$ at $t = 0$,

$$|\uparrow\rangle = \frac{1}{\sqrt{2}}(|+\rangle + |-\rangle), \tag{11.33}$$

then the evolution of the state with time governed by H_0 is given by

$$\begin{aligned} |\Psi(t)\rangle &= \frac{1}{\sqrt{2}} e^{i\omega t}|-\rangle + \frac{1}{\sqrt{2}} e^{-i\omega t}|+\rangle \\ &= \cos(\omega t)|\uparrow\rangle + \sin(\omega t)|\downarrow\rangle. \end{aligned} \tag{11.34}$$

Clearly, there is an oscillation between $|\uparrow\rangle$ and $|\downarrow\rangle$ with time. Thus an off-diagonal interaction will cause an oscillation for two degenerate states in the diagonal channel. This is the so-called Rabi oscillation and ω is called the Rabi frequency. The probability of finding state $|\uparrow\rangle$ at time t is

$$|\langle\uparrow|\Psi(t)\rangle|^2 = \cos^2(\omega t). \tag{11.35}$$

11.2.2 LANDAU-ZENER MODEL

Besides the transverse field, now we add a longitudinal Zeeman field along the z-axis, which evolves linearly with time t, namely $B_z = -bt$, with b a time-independent constant. The total Hamiltonian of this system is then defined by

$$H(t) = -g\mu_B bt\,\sigma_z + \hbar\omega\sigma_x = -\alpha t\sigma_z + \hbar\omega\sigma_x = \begin{pmatrix} -\alpha t, & \hbar\omega \\ \hbar\omega, & \alpha t \end{pmatrix}, \tag{11.36}$$

where $\alpha = g\mu_B b$ which measures the speed how the longitudinal Zeeman interaction varies with time. Eq. (11.36) is referred to as the Landau-Zener model.

The 2×2-matrix of $H(t)$ can be readily diagonalized. At a given time, its two instantaneous eigenvalues are given by

$$E_\pm(t) = \pm E(t), \qquad E(t) = \sqrt{\hbar^2\omega^2 + \alpha^2 t^2}. \tag{11.37}$$

The corresponding eigenstates are

$$|\Psi_\pm\rangle = \frac{1}{\sqrt{2E^2 \pm 2\alpha t E(t)}} \begin{pmatrix} \hbar\omega \\ \alpha t \pm E(t) \end{pmatrix}. \tag{11.38}$$

The offdiagonal term induces a gap between the two instantaneous eigenstates. The gap between $E_+(t)$ and $E_-(t)$ reaches its minimum at $t = 0$. The minimal gap is

$$\Delta = 2\hbar\omega. \tag{11.39}$$

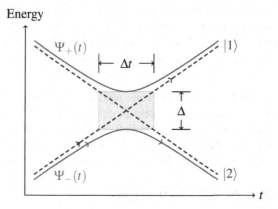

Figure 11.1 Field dependence of the diabatic (dashed lines) and adiabatic (solid curves) energies of the two-level system.

It is straightforward to show that the expectation values of ∂_t in the two instantaneous eigenstates vanish

$$\langle \Psi_- | \partial_t | \Psi_- \rangle = \langle \Psi_+ | \partial_t | \Psi_+ \rangle = 0. \tag{11.40}$$

Thus the geometric phases of these two eigenstates equal zero.

If we assume that the system starts, in the infinite past, in the lower energy eigenstate

$$|\Psi(t = -\infty)\rangle = |\Psi_-(-\infty)\rangle = |\uparrow\rangle, \tag{11.41}$$

the evolution of this quantum state depends on how fast the field evolves with time. More explicitly, the transition between $|\Psi_-\rangle$ and $|\Psi_+\rangle$ is determined by the ratio between $\langle \Psi_+ | \partial_t H | \Psi_- \rangle$ and $E_+ - E_- = 2E(t)$

$$\frac{|\langle \Psi_+ | \partial_t H | \Psi_- \rangle|}{E_+ - E_-} = \frac{\alpha \hbar \omega}{2E^2(t)}. \tag{11.42}$$

At the beginning, the gap between the two instantaneous eigenstates is significantly larger than the perturbation led by the off-diagonal term $\hbar\omega$, the chance for the particle to hop to the higher instantaneous energy eigenstate $|\Psi_+\rangle$ is small. However, with the further evolution, two limiting cases may happen:

1. Avoided level crossing: adiabatic evolution
   If the change of the field with time α is extremely small, $\alpha \to 0$, the system will evolve along the arrowed solid curve shown in Fig. 11.1, namely remain in the lower instantaneous energy eigenstate, hence $|\Psi_-(t)| \approx 1$, throughout the whole evolution process, according to the adiabatic theorem. There is no level crossing in this adiabatic evolution.
2. Complete diabatic evolution

On the other hand, if the magnetic field is switched on abruptly, $\alpha \to \infty$, the sudden approximation is valid and the system will follow the diabatic path, namely the arrowed dashed blue line in Fig. 11.1, such that the particle remains in the $|\uparrow\rangle$ state, in the whole transition from $|\Psi_1(-\infty)\rangle = |\uparrow\rangle$ to $|\Psi_2(\infty)\rangle = |\uparrow\rangle$.

In general, if the slope of the magnetic field is finite, $0 < \alpha < \infty$, there will be a finite probability of finding the system in either of the two instantaneous eigenstates. In other words, if the system is initially at the lower energy state, it has finite probability to appear in the upper energy state in the future infinite time limit. This tunneling effect is known as the Landau-Zener transition.

11.2.3   LANDAU-ZENER TRANSITION

Eq. (11.42) indicates that the Landau-Zener transition happens mainly in the time domain when the gap becomes small, namely when time is around $t = 0$. A characteristic time scale Δt within which the quantum transition becomes important is determined by demanding the change of the Zeeman energy with time, i.e. $\alpha \Delta t$, equal to the gap at $t = 0$, namely

$$\Delta t = \frac{\Delta}{\alpha} = \frac{2\hbar\omega}{\alpha}. \tag{11.43}$$

In the phase space, this transition important region is schematically shown by the light green rectangle in Fig. 11.1. The area of this light green rectangle is

$$S = \Delta \times \Delta t = \frac{4\hbar^2\omega^2}{\alpha}. \tag{11.44}$$

Not surprisingly, the transition probability from the lower to higher energy state is determined by the size of $S/\hbar$. The higher is $S/\hbar$, the lower is the transition rate. It has been shown by Lev Landau, and later independently by Clarence Zener, that the probability for the particle to transit from the lower energy state from infinite past to the upper energy state at infinite future is

$$\boxed{P_{LZ} = \exp\left(-\frac{\pi\hbar\omega^2}{\alpha}\right) = \exp\left(-\frac{\pi S}{4\hbar}\right).} \tag{11.45}$$

A derivation of this formula is given below.

Since ω is inversely proportional to the periodic time of the Rabi oscillation, τ_r

$$\tau_r = \frac{2\pi}{\omega}, \tag{11.46}$$

the above probability can be also represented using the ratio between Δt and τ_r as

$$P_{LZ} = \exp\left(-\pi^2 \frac{\Delta t}{\tau_r}\right). \tag{11.47}$$

The adiabatic process corresponds to the limit the energy evolution time Δt is significantly larger than the Rabi oscillation time τ_r, i.e. $\Delta t \gg \tau_r$ or $\hbar\omega^2 \gg \alpha$. On the other hand, if the Rabi oscillation time is much larger than the energy evolution time, $\tau_r \gg \Delta t$ or $\hbar\omega^2 \ll \alpha$, the tunneling is completely transparent and the transition rate from $\Psi_-(-\infty)$ to $\Psi_+(\infty)$ is 1.

11.2.4 DERIVATION OF THE LANDAU-ZENER FORMULAS*

The Landau-Zener model can be rigorously solved. Below we derive the Landau-Zener formula (11.45) using an approach introduced by Ho and Chibotaru.[1] We expand the quantum state in the representation of σ_z eigenstates

$$|\Psi(t)\rangle = A(t)|\uparrow\rangle + B(t)|\downarrow\rangle. \tag{11.48}$$

The Landau-Zener formulas is to calculate the transition probability from an instantaneous lower energy eigenstate at infinite past, i.e. $|\Psi_-(-\infty)\rangle = |\uparrow\rangle$, to an instantaneous upper energy eigenstate at infinite future, i.e. $|\Psi_+(\infty)\rangle = |\uparrow\rangle$. As both the initial and final states are in the up-spin states, we just need to calculate the ratio $|A(\infty)/A(-\infty)|^2$.

Substituting the above wave function into the time-dependent Schrödinger equation

$$i\hbar\frac{\partial}{\partial t}|\Psi(t)\rangle = H(t)|\Psi(t)\rangle \tag{11.49}$$

yields two coupled equations

$$i\hbar\partial_t A = -\alpha t A + \hbar\omega B, \tag{11.50}$$
$$i\hbar\partial_t B = \alpha t B + \hbar\omega A. \tag{11.51}$$

Eliminating B from the above equation, we obtain the differential equation for A

$$\hbar^2\partial_t^2 A + (E^2 - i\alpha\hbar)A = 0. \tag{11.52}$$

In the limit $|t| \to \infty$, the phase of $A(t)$ oscillates quickly with time, but its modulus should converge to a time-independent constant. Thus we can write $A(t)$ in this limit as

$$A(t) = |A|e^{i\phi(t)}, \tag{11.53}$$

where $|A|$ is the modulus of $A(t)$, which is time-independent in the limit $t \to \pm\infty$. Substituting the above expression into Eq. (11.52) yields

$$i\hbar^2\partial_t^2\phi - \hbar^2(\partial_t\phi)^2 + E^2 - i\alpha\hbar = 0. \tag{11.54}$$

Separating real and imaginary parts gives

$$\partial_t\phi = \pm\frac{E}{\hbar}, \tag{11.55}$$
$$\partial_t^2\phi = \frac{\alpha}{\hbar}. \tag{11.56}$$

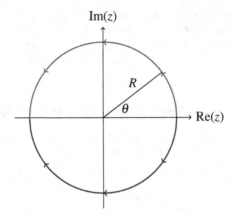

Figure 11.2   Upper and lower half circle contours used in the Cauchy integration (11.61).

The solution of $\partial_t \phi$ that satisfies the above two equations is

$$
\partial_t \phi =
\begin{cases}
\dfrac{E}{\hbar}, & t \to \infty, \\[2ex]
-\dfrac{E}{\hbar}, & t \to -\infty.
\end{cases}
\tag{11.57}
$$

Hence in the limit $t \to \pm\infty$

$$
\partial_t \phi = \frac{\alpha t}{\hbar} \sqrt{1 + \left(\frac{\hbar\omega}{\alpha t}\right)^2} \approx \frac{\alpha t}{\hbar} + \frac{\hbar\omega^2}{2\alpha t}.
\tag{11.58}
$$

The ratio between $\partial_t A$ and A is then found to be

$$
\left(\frac{\partial_t A}{A}\right)_{t \to \pm\infty} = i(\partial_t \phi)_{t \to \pm\infty} \approx i\left(\frac{\alpha t}{\hbar} + \frac{\hbar\omega^2}{2\alpha t}\right),
\tag{11.59}
$$

which is an analytic function of t that can be extended from the real axis to the complex plane. This allows us to calculate the integral

$$
\int_{-\infty}^{\infty} dt\, \frac{\partial_t A}{A} = \ln \frac{A(\infty)}{A(-\infty)}
\tag{11.60}
$$

using the Cauchy integral theorem

$$
\int_{-\infty}^{\infty} dt\, \frac{\partial_t A}{A}(t) = -\int_C dt\, \frac{\partial_t A}{A}(z)
\tag{11.61}
$$

---

[1]L. T. A. Ho and L. F. Chibotaru, Phys. Chem. Chem.Phys. **16**, 6942 (2014)

on an infinite semi-circle C centered around $z = 0$ on the upper or lower half complex plane, shown in Fig. 11.2, without invoking the exact solution of $A(t)$ on the whole time domain. The contour integration is done by setting $z = R\exp(i\theta)$,

$$
\begin{aligned}
-\int_C dt \frac{\partial_t A}{A}(z) &= -i \int_C dz \left(\frac{\alpha z}{\hbar} + \frac{\hbar \omega^2}{2\alpha z} \right) \\
&= \int_0^{\pm \pi} d\theta \left(\frac{\alpha R^2 e^{2i\theta}}{\hbar} + \frac{\hbar \omega^2}{2\alpha} \right) \\
&= \pm \frac{\pi \hbar \omega^2}{2\alpha},
\end{aligned}
\tag{11.62}
$$

where $\pm$ corresponds to the contour on the upper and lower half plane, respectively. Therefore, we have

$$
A(\infty) = \exp \left(\pm \frac{\pi \hbar \omega^2}{2\alpha} \right) A(-\infty),
\tag{11.63}
$$

and

$$
|A(\infty)|^2 = \exp \left(\pm \frac{\pi \hbar \omega^2}{\alpha} \right) |A(-\infty)|^2.
\tag{11.64}
$$

Thus the transition probability from the lower energy eigenstate at infinite past, i.e $A(-\infty) = 1$, to the upper energy eigenstate at infinite future is

$$
P_{LZ} = \exp \left(-\frac{\pi \hbar \omega^2}{\alpha} \right).
\tag{11.65}
$$

The positive solution in (11.64) is discarded because the probability cannot be larger than 1.

11.3 BERRY'S PHASE

Eq. (11.14) shows that a particle which is initially in an eigenstate of the Hamiltonian will remain in that eigenstate under an adiabatic evolution and just acquire a phase factor in its eigenfunction. This phase contains two parts, one is the dynamic phase and the other is the geometric phase. If the time evolution of the Hamiltonian is due to the change of some parameters $\mathbf{R}(t)$, then

$$
\frac{\partial}{\partial t} |m\rangle = \frac{\partial \mathbf{R}}{\partial t} \cdot \nabla_{\mathbf{R}} |m\rangle
\tag{11.66}
$$

and the geometric phase becomes

$$
\begin{aligned}
\gamma_n(t) &= i \int_0^t \langle n | \nabla_{\mathbf{R}} | n \rangle \cdot \frac{\partial \mathbf{R}}{\partial t} dt \\
&= i \int_{\mathbf{R}(0)}^{\mathbf{R}(t)} \langle n | \nabla_{\mathbf{R}} | n \rangle \cdot d\mathbf{R}.
\end{aligned}
\tag{11.67}
$$

This is a path integral along the curve defined by the variance of $\mathbf{R}(t)$ with time. $d\mathbf{R}$ is a tangent vector along the path.

If $\mathbf{R(t)}$ returns to its original value along a path C after a time T, the net geometric phase is simply a line integral along that close path

$$\boxed{\gamma_n(t) = i \oint_C \langle n | \nabla_{\mathbf{R}} | n \rangle \cdot d\mathbf{R}.} \tag{11.68}$$

This integral is generally not zero. This expression of the geometric phase is also called Berry's phase. The Berry's phase depends only on the path taken, not on how fast the particle travels, provided the adiabatic condition is valid.

The phenomenon of Berry phase was first discovered by Pancharantnam in 1956, independently by Longuet-Higgins in 1958, and more generally by Berry in 1984.

11.3.1 FICTITIOUS GAUGE VECTOR

By defining a vector gauge potential, which is also refereed as the Berry connection,

$$\boxed{\mathscr{A}_n(\mathbf{R}) = i \langle n | \nabla_{\mathbf{R}} | n \rangle,} \tag{11.69}$$

the geometric phase can be expressed as

$$\boxed{\gamma_n(t) = \oint_C \mathscr{A}_n(\mathbf{R}) \cdot d\mathbf{R}.} \tag{11.70}$$

By contrast, the dynamic phase depends strictly on the elapsed time

$$\theta_n(t) = -\frac{1}{\hbar} \int_0^T E_n(t) dt. \tag{11.71}$$

11.3.2 FICTITIOUS MAGNETIC FIELD

Using Stokes' theorem, the line integral of the Berry's phase can be written as an area integral if we introduce a fictitious magnetic field $\mathbf{B}_n$

$$\boxed{\mathscr{B}_n(\mathbf{R}) = \nabla_{\mathbf{R}} \times \mathscr{A}_n(\mathbf{R}) = i \nabla_{\mathbf{R}} \times \langle n | \nabla_{\mathbf{R}} | n \rangle.} \tag{11.72}$$

Both $\mathscr{A}_n(\mathbf{R})$ and $\mathscr{B}_n(\mathbf{R})$ are real quantities. Now the Berry's phase can be expressed as

$$\boxed{\gamma_n(t) = \int_S \mathscr{B}_n(\mathbf{R}) \cdot d\mathbf{a},} \tag{11.73}$$

which is just the fictitious flux penetrating the area S enclosed by C.

Since the curl of a curl vanishes and $(\nabla_{\mathbf{R}})^\dagger = -\nabla_{\mathbf{R}}$, the fictitious field can be also expressed as

$$\begin{aligned} \mathscr{B}_n(\mathbf{R}) &= i \langle \nabla_{\mathbf{R}} n | \times \nabla_{\mathbf{R}} | n \rangle \\ &= -\sum_m i \langle n | \nabla_{\mathbf{R}} | m \rangle \times \langle m | \nabla_{\mathbf{R}} | n \rangle \\ &= -\sum_{m \neq n} i \langle n | \nabla_{\mathbf{R}} | m \rangle \times \langle m | \nabla_{\mathbf{R}} | n \rangle. \end{aligned} \tag{11.74}$$

In this equation, the term with $m = n$ is discarded because

$$\langle n|\nabla_{\mathbf{R}}|n\rangle \times \langle n|\nabla_{\mathbf{R}}|n\rangle = 0. \qquad (11.75)$$

By taking the $\mathbf{R}$-gradient for the eigen-equation

$$H|n\rangle = E_n|n\rangle, \qquad (11.76)$$

we get

$$(\nabla_{\mathbf{R}}H)|n\rangle + H\nabla_{\mathbf{R}}|n\rangle = (\nabla_{\mathbf{R}}E_n)|n\rangle + E_n\nabla_{\mathbf{R}}|n\rangle. \qquad (11.77)$$

By further taking the inner product with $\langle\Psi_m|$ $(m \neq n)$

$$\langle m|(\nabla_{\mathbf{R}}H)|n\rangle + \langle m|H\nabla_{\mathbf{R}}|n\rangle = E_n\langle m|\nabla_{\mathbf{R}}|n\rangle. \qquad (11.78)$$

Finally we find that

$$\langle m|\nabla_{\mathbf{R}}|n\rangle = \frac{\langle m|(\nabla_{\mathbf{R}}H)|n\rangle}{E_n - E_m}. \qquad (11.79)$$

This enables us to write

$$\boxed{\mathscr{B}_n(\mathbf{R}) = i\sum_{m\neq n} \frac{\langle n|(\nabla_{\mathbf{R}}H)|m\rangle \times \langle m|(\nabla_{\mathbf{R}}H)|n\rangle}{(E_n - E_m)^2}.} \qquad (11.80)$$

11.3.3 QUANTIZATION OF FICTITIOUS MAGNETIC FLUX

Consider two surfaces S_1 and S_2 in $\mathbf{R}$-space, each bounded by the same curve. The Berry phase results from each of these two surfaces should differ only by an integer multiple of 2π

$$\int_{S_1} d\mathbf{a} \cdot \mathscr{B}_n - \int_{S_2} d\mathbf{a} \cdot \mathscr{B}_n = 2\pi k, \qquad (11.81)$$

where k is an integer. The left hand side is actually the total flux penetrating the surface enclosed by S_1 and S_2, we thus have

$$\boxed{\oint_{S_1 \oplus S_2} d\mathbf{a} \cdot \mathscr{B}_n = 2\pi k,} \qquad (11.82)$$

which means that the fictitious magnetic field in a closed surface must be quantized.

11.3.4 A SPIN-1/2 PARTICLE IN A MAGNETIC FLUX

As an example, let us calculate the Berry phase for a spin-1/2 particle manipulated slowly through a time-varying magnetic field $\mathbf{R}(t)$. The Hamiltonian reads

$$H = -\mu\sigma \cdot \mathbf{R}(t), \qquad (11.83)$$

where μ is the magnetic moment of this particle.

In the reference frame where the z-axis points along the direction of $\mathbf{R}(t)$,

$$
\begin{aligned}
\sigma &= \frac{1}{2}\left(\sigma^+ + \sigma^-\right)\hat{x} + \frac{1}{2i}\left(\sigma^+ - \sigma^-\right)\hat{y} + \sigma_z\hat{z} \\
&= \frac{1}{2}(\hat{x} - i\hat{y})\sigma^+ + \frac{1}{2}(\hat{x} + i\hat{y})\sigma^- + \sigma_z\hat{z}
\end{aligned}
$$

and the up and down spin states have respectively the energies

$$
E_\pm = \mp \mu R(t), \tag{11.84}
$$

where $R(t)$ is the magnitude of the magnetic field. The matrix elements of the spin operators between these two eigenstates are

$$
\begin{aligned}
\langle\Psi_-|\sigma^+|\Psi_+\rangle &= \langle\Psi_-|\sigma_z|\Psi_+\rangle = 0, \tag{11.85} \\
\langle\Psi_-|\sigma^-|\Psi_+\rangle &= 2. \tag{11.86}
\end{aligned}
$$

Since the $\mathbf{R}$-derivative of the Hamiltonian is

$$
\nabla_{\mathbf{R}}H = -\mu\sigma, \tag{11.87}
$$

we therefore have

$$
\begin{aligned}
\mathscr{B}_+(\mathbf{R}) &= i\frac{\langle\Psi_+|(\nabla_{\mathbf{R}}H)|\Psi_-\rangle\langle\Psi_-|(\nabla_{\mathbf{R}}H)|\Psi_+\rangle}{(E_+ - E_-)^2} \\
&= i\frac{\langle\Psi_+|(\hat{x} - i\hat{y})\sigma^+|\Psi_-\rangle \times \langle\Psi_-|(\hat{x} + i\hat{y})\sigma^-|\Psi_+\rangle}{16R^2} \\
&= -\frac{1}{2R^2}\hat{\mathbf{R}}. \tag{11.88}
\end{aligned}
$$

Similarly, we have

$$
\mathscr{B}_-(\mathbf{R}) = i\frac{\langle\Psi_-|(\nabla_{\mathbf{R}}H)|\Psi_+\rangle\langle\Psi_+|(\nabla_{\mathbf{R}}H)|\Psi_-\rangle}{(E_+ - E_-)^2} = \frac{1}{2R^2}\hat{\mathbf{R}}. \tag{11.89}
$$

Finally, we calculate Berry's phase to be

$$
\gamma_\pm(t) = \int_S \mathscr{B}_\pm(\mathbf{R})\cdot d\mathbf{a} = \mp\frac{1}{2}\int_S \frac{1}{R^2}\hat{\mathbf{R}}\cdot d\mathbf{a} = \mp\frac{1}{2}\Omega, \tag{11.90}
$$

where Ω is the solid angle subtended by the path through which the magnetic field travels relative to the origin $\mathbf{R} = 0$.

$\gamma_\pm$ does not depend on the magnetic moment μ of the particle. However, the sign of μ determines the direction of the "up" spin, which implies that from the sign of $\gamma_\pm(t)$, one can determine the sign of μ.

11.3.5 CHARGED PARTICLE MOVING AROUND A MAGNETIC FLUX

In Section 4.3.4, we show that a charged particle moving along a closed path that circulates a solenoid carrying a steady current but in the region in which the magnetic field is absent acquires a phase determined purely by the magnetic flux penetrating the solenoid. This phase could be detected by measuring the interference effect proposed by Aharonov and Bohm. We now show that this phase is also the Barry phase of the particle accumulated along that closed path.

For simplicity, we consider a circular path, with a radius of R. The solenoid is at the center of the circle. The Hamiltonian is defined by

$$H = \frac{1}{2\mu}\left(-i\hbar\nabla - e\mathbf{A}\right)^2 = \frac{1}{2\mu}\left(-\frac{i\hbar}{R}\frac{\partial}{\partial\phi} - \frac{e\Phi}{2\pi R}\right)^2, \qquad (11.91)$$

where $\mathbf{A}$ is the gauge potential generated by the magnetic flux Φ at the center

$$\mathbf{A} = \frac{\Phi}{2\pi R}\hat{\phi}. \qquad (11.92)$$

The eigenfunction of this Hamiltonian has the form

$$\Psi(\phi) = \frac{1}{\sqrt{2\pi}}e^{in\phi}, \qquad (11.93)$$

where n is an integer since $\Phi(\phi)$ is a periodic function of ϕ and $\Phi(2\pi) = \Phi(0)$. The eigenenergy is

$$E = \frac{\hbar^2}{2\mu R^2}\left(n - \frac{e\Phi}{2\pi\hbar}\right)^2. \qquad (11.94)$$

Clearly, the eigenenergy depends on the magnetic field inside the solenoid, even though the field on the circle is zero. Again it indicates that it is the gauge potential, not the magnetic field $\mathbf{B}$, that is physically more fundamental.

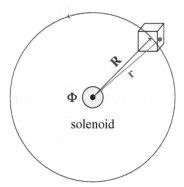

To calculate the Berry phase, we put the particle inside a small box and move the box to circulate the solenoid. The instantaneous Hamiltonian of the particle in the box centered at $\mathbf{R}$ is

$$H = \frac{1}{2\mu}\left(-i\hbar\nabla - e\mathbf{A}\right)^2 + V(\mathbf{r} - \mathbf{R}), \qquad (11.95)$$

where $V(\mathbf{r} - \mathbf{R})$ is the confined box potential. $\mathbf{R}$ defines the parameter path that is used for evaluating the Berry phase.

As $\mathbf{A}$ is a curl-free vector potential, we can take a gauge transformation to eliminate the vector potential term in the Hamiltonian

$$\Psi(\mathbf{r}) = e^{i\gamma(\mathbf{r})}\tilde{\Psi}(\mathbf{r} - \mathbf{R}), \tag{11.96}$$

where

$$\gamma(\mathbf{r}) = \frac{e}{\hbar}\int_{\mathbf{R}}^{\mathbf{r}} \mathbf{A}(\mathbf{r}) \cdot d\mathbf{r}. \tag{11.97}$$

$\tilde{\Psi}(\mathbf{r} - \mathbf{R})$ is the eigenfunction of the particle inside the box without gauge potential, determined by the eigen-equation

$$\left[-\frac{\hbar^2}{2\mu}\nabla^2 + V(\mathbf{r} - \mathbf{R}) \right] \tilde{\Psi}(\mathbf{r} - \mathbf{R}) = E\tilde{\Psi}(\mathbf{r} - \mathbf{R}). \tag{11.98}$$

We thus have

$$\nabla_{\mathbf{R}}\Psi = \nabla_{\mathbf{R}}\left[e^{i\gamma(\mathbf{r})}\tilde{\Psi}(\mathbf{r} - \mathbf{R}) \right] = -\frac{ie}{\hbar}\mathbf{A}(\mathbf{R})\Psi + e^{i\gamma(\mathbf{r})}\nabla_{\mathbf{R}}\tilde{\Psi}(\mathbf{r} - \mathbf{R}), \tag{11.99}$$

and

$$\langle\Psi|\nabla_{\mathbf{R}}|\Psi\rangle = -\frac{ie}{\hbar}\mathbf{A}(\mathbf{R}) - \langle\tilde{\Psi}(\mathbf{r} - \mathbf{R})|\nabla_{\mathbf{r}}|\tilde{\Psi}(\mathbf{r} - \mathbf{R})\rangle. \tag{11.100}$$

Since the particle is confined in the box, the expectation value of the momentum operator is zero for all the eigenstates

$$\langle\tilde{\Psi}(\mathbf{r} - \mathbf{R})|\nabla_{\mathbf{r}}|\tilde{\Psi}(\mathbf{r} - \mathbf{R})\rangle = 0. \tag{11.101}$$

Therefore, we have

$$\langle\Psi|\nabla_{\mathbf{R}}|\Psi\rangle = -\frac{ie}{\hbar}\mathbf{A}(\mathbf{R}). \tag{11.102}$$

The Berry phase is

$$\gamma = \frac{e}{\hbar}\oint_C \mathbf{A}(\mathbf{R}) \cdot d\mathbf{R} = \frac{e\Phi}{\hbar}, \tag{11.103}$$

which is just the phase obtained from the Aharonov-Bohm effect, shown in Eq. (4.79).

11.4 PROBLEMS

1. A particle of mass μ is bound to a delta-function potential

$$V = -\alpha\delta(x) \qquad (a>0).$$

   a. The strength α is suddenly doubled, i.e. $\alpha \to 2\alpha$. What is the probability that the particle remains bound?

   b. Calculate the geometric phase change when α gradually increases from α to 2α.

   c. If the increase occurs at a constant rate, $d\alpha/dt = v$, what is the dynamic phase change for this process?

2. a. Show that the fictitious magnetic field $\mathscr{B}(\mathbf{R}) = i\nabla_{\mathbf{R}} \times \langle n| \nabla_{\mathbf{R}}|n\rangle$ is invariant under the gauge transformation $|n\rangle \to e^{i\Lambda(\mathbf{R})}|n\rangle$.

   b. If the fictitious magnetic field enclosed in a sphere is

$$\mathscr{B}(\mathbf{r}) = \frac{\hat{\mathbf{r}}}{2r^2},$$

what is the total flux penetrating this sphere?

3. a. Calculate the geometric phase change when the one-dimensional infinite square well expands adiabatically from width L to $2L$.

   b. If the expansion occurs at a constant rate, $dL/dt = a$, what is the dynamic phase change for this process?

   c. If the well now contracts back to its original size, what is Berry's phase for this cycle?

4. Show that if the instantaneous eigenvector $\Psi_n(\mathbf{r},t) = \langle\mathbf{r}|n(t)\rangle$ is real, the geometric phase vanishes.

12 Relativistic Quantum Mechanics

12.1 RELATIVISTIC COVARIANCE

In special relativity, time and space are transformed according to the Lorentz transformation

$$t' = \frac{c}{\sqrt{c^2 - v^2}}\left(t - \frac{vx}{c^2}\right), \tag{12.1}$$

$$x' = \frac{c}{\sqrt{c^2 - v^2}}(x - vt), \tag{12.2}$$

$$y' = y, \tag{12.3}$$

$$z' = z. \tag{12.4}$$

This transformation keeps the space-time scalar

$$c^2 t'^2 - x'^2 - y'^2 - z'^2 = c^2 t^2 - x^2 - y^2 - z^2 \tag{12.5}$$

invariant.

The transformation for the four-velocity vector is

$$\frac{\partial}{\partial t} = \frac{c}{\sqrt{c^2 - v^2}}\left(\frac{\partial}{\partial t'} - v\frac{\partial}{\partial x'}\right), \tag{12.6}$$

$$\frac{\partial}{\partial x} = \frac{c}{\sqrt{c^2 - v^2}}\left(-\frac{v}{c^2}\frac{\partial}{\partial t'} + \frac{\partial}{\partial x'}\right) \tag{12.7}$$

$$\frac{\partial}{\partial y} = \frac{\partial}{\partial y'}, \tag{12.8}$$

$$\frac{\partial}{\partial z} = \frac{\partial}{\partial z'}. \tag{12.9}$$

It also leaves the four-velocity scalar

$$\left(\frac{\partial}{\partial t}\right)^2 - c^2\left(\frac{\partial}{\partial x}\right)^2 - c^2\left(\frac{\partial}{\partial y}\right)^2 - c^2\left(\frac{\partial}{\partial z}\right)^2$$

$$= \left(\frac{\partial}{\partial t'}\right)^2 - c^2\left(\frac{\partial}{\partial x'}\right)^2 - c^2\left(\frac{\partial}{\partial y'}\right)^2 - c^2\left(\frac{\partial}{\partial z'}\right)^2 \tag{12.10}$$

unchanged under the transformation.

However, the Schrödinger equation

$$i\hbar\frac{\partial}{\partial t}\Psi(\mathbf{r},t) = \left[-\frac{\hbar^2}{2\mu}\nabla^2 + V(\mathbf{r})\right]\Psi(\mathbf{r},t) \tag{12.11}$$

DOI: 10.1201/9781003174882-12

is not relativistically covariant because the time derivative and the spatial derivative appear in different orders.

In special relativity, the energy of a free particle is given by

$$E = \sqrt{\mu^2 c^4 + c^2 \mathbf{p}^2}. \tag{12.12}$$

To fulfill the relativistic covariance, it seems that one can define the Hamiltonian as

$$H = \sqrt{\mu^2 c^4 + c^2 \mathbf{p}^2} = \mu c^2 + \frac{\mathbf{p}^2}{2\mu} - \frac{\mathbf{p}^4}{8\mu^3 c^2} + \cdots. \tag{12.13}$$

However, it has some serious shortcomings:

1. It is impossible to formulate a covariant wave equation, because time and space are not treated on an equal footing. We have one time derivative and an infinite series of increasing spatial derivatives from the momentum operator.
2. The wave equation is local in time, but nonlocal in space, again due to the presence of infinite spatial derivatives, hence violating the principle of causality.

Relativistic quantum mechanics is quantum mechanics that is consistent with special relativity. Nevertheless, relativistic quantum mechanics is only an approximation to a fully self-consistent relativistic theory of known particle interactions because it does not describe cases where the number of particles changes. In special relativity, mass (particle) can emerge from energy, which is not possible in quantum mechanics based on the conservation of probability.

12.2   KLEIN-GORDON EQUATION

The Klein-Gordon equation applies to spinless particles such as mesons. It is obtained by taking the time derivative one more time such that

$$-\hbar^2 \frac{\partial^2}{\partial t^2} \Psi = i\hbar \frac{\partial}{\partial t} H\Psi = H^2 \Psi. \tag{12.14}$$

Taking

$$H^2 = \mu^2 c^4 - \hbar^2 c^2 \nabla^2, \tag{12.15}$$

we then obtain the Klein-Gordon equation

$$-\hbar^2 \frac{\partial^2}{\partial t^2} \Psi = \left(\mu^2 c^4 - \hbar^2 c^2 \nabla^2 \right) \Psi. \tag{12.16}$$

It looks like a classical wave equation, except for the mass term.

The Klein-Gordon equation is relativistically covariant. An obvious solution of this equation is

$$\Psi_{\mathbf{p},E}(\mathbf{r},t) = \exp\left(\frac{i\mathbf{p} \cdot \mathbf{r} - iEt}{\hbar} \right) \tag{12.17}$$

where $\mathbf{p}$ is the momentum and E is the energy. This wave function gives the correct relativistic dispersion for a free particle

$$E^2 = \mu^2 c^4 + c^2 \mathbf{p}^2. \tag{12.18}$$

12.2.1 CURRENT AND DENSITY OPERATORS

Multiplying Eq. (12.16) from the left by Ψ^*, we have

$$-\hbar^2 \Psi^* \frac{\partial^2}{\partial t^2} \Psi = \Psi^* \left(\mu^2 c^4 - \hbar^2 c^2 \nabla^2 \right) \Psi. \tag{12.19}$$

Taking the complex conjugate for both sides of the equation, we further obtain

$$-\hbar^2 \Psi \frac{\partial^2}{\partial t^2} \Psi^* = \Psi \left(\mu^2 c^4 - \hbar^2 c^2 \nabla^2 \right) \Psi^*. \tag{12.20}$$

Subtract these two equations, the terms without derivatives are cancelled, which leads to

$$\Psi^* \frac{\partial^2}{\partial t^2} \Psi - \Psi \frac{\partial^2}{\partial t^2} \Psi^* = c^2 \left(\Psi^* \nabla^2 \Psi - \Psi \nabla^2 \Psi^* \right). \tag{12.21}$$

It can be further written as

$$\frac{1}{c^2} \partial_t \left(\Psi^* \partial_t \Psi - \Psi \partial_t \Psi^* \right) - \nabla \cdot \left(\Psi^* \nabla \Psi - \Psi \nabla \Psi^* \right) = 0. \tag{12.22}$$

This is the equation of continuity, or the equation of "charge" conservation. It can be also expressed as

$$\partial_t \rho + \nabla \cdot \mathbf{j} = 0, \tag{12.23}$$

where ρ and $\mathbf{j}$ are the charge density and current operators, respectively

$$\rho = i \frac{\hbar}{2\mu c^2} \left(\Psi^* \partial_t \Psi - \Psi \partial_t \Psi^* \right), \tag{12.24}$$

$$\mathbf{j} = -i \frac{\hbar}{2\mu} \left(\Psi^* \nabla \Psi - \Psi \nabla \Psi^* \right). \tag{12.25}$$

The probability current operator has exactly the same form as in nonrelativistic quantum mechanics. But the definition of charge density operator is changed. In nonrelativistic quantum mechanics, the probability density operator is defined by $\rho = |\Psi|^2$.

The probability density such defined is conserved. However, this density operator is not positive definite.

12.2.2 INTERPRETATION OF THE DENSITY OPERATOR

Unlike the Schrödinger equation, the Klein-Gordon equation is second order in time derivatives. That means one should specify the values of both the wave function and its first derivation at zero time in order to find its solution.

To understand this, let us reduce the second-order Klein-Gordon equation to two first-order equations by defining two new variables

$$\theta = \frac{1}{2}\left(\Psi + \frac{i\hbar}{\mu c^2}\partial_t\Psi\right), \tag{12.26}$$

$$\phi = \frac{1}{2}\left(\Psi - \frac{i\hbar}{\mu c^2}\partial_t\Psi\right). \tag{12.27}$$

Clearly,

$$\theta + \phi = \Psi, \tag{12.28}$$

$$\theta - \phi = \frac{i\hbar}{\mu c^2}\partial_t\Psi. \tag{12.29}$$

Instead of specifying Ψ and $\partial_t\Psi$ at $t = 0$, we can just specify the values of θ and ϕ at $t = 0$. Using these two variables, the Klein-Gordon equation can be written as two coupled equations

$$i\hbar\partial_t(\theta - \phi) = -\frac{\hbar^2}{\mu c^2}\partial_t^2\Psi = \mu c^2(\theta + \phi) - \frac{\hbar^2\nabla^2}{\mu}(\theta + \phi), \tag{12.30}$$

$$i\hbar\partial_t(\theta + \phi) = \mu c^2(\theta - \phi). \tag{12.31}$$

They can be also represented as

$$i\hbar\partial_t\theta = -\frac{\hbar^2\nabla^2}{2\mu}(\theta + \phi) + \mu c^2\theta, \tag{12.32}$$

$$i\hbar\partial_t\phi = \frac{\hbar^2\nabla^2}{2\mu}(\theta + \phi) - \mu c^2\phi. \tag{12.33}$$

If we define

$$\chi = \begin{pmatrix} \theta \\ \phi \end{pmatrix}, \tag{12.34}$$

the above equations become

$$i\hbar\partial_t\chi = -\frac{\hbar^2\nabla^2}{2\mu}(\sigma_3 + i\sigma_2)\chi + \mu c^2\sigma_3\chi, \tag{12.35}$$

where σ_2 and σ_3 are Pauli matrices.

Now the density operator becomes

$$\begin{aligned} \rho &= i\frac{\hbar}{2\mu c^2}(\Psi^*\partial_t\Psi - \Psi\partial_t\Psi^*) \\ &= \frac{1}{2}(\theta^* + \phi^*)(\theta - \phi) + \frac{e}{2}(\theta + \phi)(\theta^* - \phi^*) \\ &= \theta^*\theta - \phi^*\phi. \end{aligned} \tag{12.36}$$

ρ is diagonal in this (θ,ϕ) representation. However, the probability current operator

$$\mathbf{j} = -i\frac{\hbar}{2\mu}\left[(\theta^* + \phi^*)\nabla(\theta + \phi) - (\theta + \phi)\nabla(\theta^* + \phi^*)\right],\qquad(12.37)$$

is not diagonal in this representation.

If we define the charge density by multiplying ρ with a charge parameter e

$$\rho_e = e\rho = e(\theta^*\theta - \phi^*\phi),\qquad(12.38)$$

we may interpret ρ_e as a probability charge density operator. θ is the wave function of a positive charge, and ϕ is that of a negative charge.

12.2.3 NEGATIVE ENERGY

Eq. (12.18) implies that there are both positive and negative energy solutions. These solutions can be more clearly seen by directly solving Eq. (12.35). To set

$$\chi(\mathbf{r},t) = \chi_0 \exp\left(\frac{i\mathbf{p}\cdot\mathbf{r} - iEt}{\hbar}\right),\qquad(12.39)$$

we have

$$
\begin{aligned}
E\chi_0 &= \frac{\mathbf{p}^2}{2\mu}(\sigma_3 + i\sigma_2) + \mu c^2\sigma_3\chi_0 \\
&= \begin{pmatrix} \dfrac{\mathbf{p}^2}{2\mu} + \mu c^2 & \dfrac{\mathbf{p}^2}{2\mu} \\[2mm] \dfrac{\mathbf{p}^2}{-2\mu} & -\dfrac{\mathbf{p}^2}{2\mu} - \mu c^2 \end{pmatrix}\chi_0.
\end{aligned}\qquad(12.40)
$$

The eigenenergy is given by $E = \pm E_\mathbf{p}$ with

$$E_\mathbf{p} = \sqrt{\mathbf{p}^2 c^2 + \mu^2 c^4}.\qquad(12.41)$$

$\chi_0 = (\theta_0, \phi_0)^T$ are determined by

$$\left(E - \mu c^2\right)\theta_0 - \left(E + \mu c^2\right)\phi_0 = 0.\qquad(12.42)$$

We therefore have

$$\chi_+ = \begin{pmatrix} E_\mathbf{p} + \mu c^2 \\ E_\mathbf{p} - \mu c^2 \end{pmatrix}\exp\left(\frac{i\mathbf{p}\cdot\mathbf{r} - iE_\mathbf{p}t}{\hbar}\right)\qquad(12.43)$$

if $E = E_\mathbf{p}$, and

$$\chi_- = \begin{pmatrix} E_\mathbf{p} - \mu c^2 \\ E_\mathbf{p} + \mu c^2 \end{pmatrix}\exp\left(\frac{i\mathbf{p}\cdot\mathbf{r} + iE_\mathbf{p}t}{\hbar}\right)\qquad(12.44)$$

if $E = -E_\mathbf{p}$.

Thus either positive or negative energy states have the contribution of both positive and negative charged particles. Only when the particle is at rest, i.e. when $E = \pm\mu c^2$, the positive energy has only the contribution from the positive charge, and the negative energy has only the contribution from the negative charge.

12.2.4 NONRELATIVISTIC LIMIT

Let us consider the nonrelativistic limit of the Klein-Gordon equation, and see if it can be reduced to the Schrödinger equation. To recover the nonrelativistic limit, we set the wave function as

$$\Psi(\mathbf{r},t) = \psi(\mathbf{r},t)\exp\left(-i\mu c^2 t/\hbar\right). \tag{12.45}$$

Substituting it into the Klein-Gordon equation, we obtain

$$i\hbar\frac{\partial}{\partial t}\psi - \frac{\hbar^2}{2\mu c^2}\frac{\partial^2}{\partial t^2}\psi = -\frac{\hbar^2}{2\mu}\nabla^2\psi. \tag{12.46}$$

The second term on the left contains the second time derivative divided by μc^2, which is negligible in the nonrelativistic limit. Thus the Klein-Gordon equation reduces to the Schrödinger equation in the nonrelativistic limit,

$$i\hbar\frac{\partial}{\partial t}\psi \approx -\frac{\hbar^2}{2\mu}\nabla^2\psi. \tag{12.47}$$

Similarly, we can write the probability density operator as

$$
\begin{aligned}
\rho &= i\frac{\hbar}{2\mu c^2}\left(\Psi^*\partial_t\Psi - \Psi\partial_t\Psi^*\right) \\
&= \frac{i\hbar}{2\mu c^2}\left(\psi^*\partial_t\psi - \psi\partial_t\psi^*\right) + e|\psi|^2.
\end{aligned} \tag{12.48}
$$

On the right, the first term in the second line can be neglected, and the density operator becomes

$$\rho \approx |\psi|^2 \tag{12.49}$$

in the nonrelativistic limit. This recovers the expression of the probability density in the nonrelativistic limit.

The current operator is unchanged in the nonrelativistic limit since it has no time derivatives.

12.3 DIRAC EQUATION

The Dirac equation applies to spin-1/2 fermions, such as electrons, neutrinos, or nucleons. From the Dirac equation, the relativistic quantum mechanics predicts successfully the existence of antimatter, electron spin, internal magnetic moments of elementary spin 1/2 fermions, fine structure, and quantum dynamics of charged particles in electromagnetic fields.

12.3.1 "DERIVATION" OF THE DIRAC EQUATION

The Dirac equation is a gift of nature. It cannot be derived from microscopic principle.

Dirac proposed the equation named after him to solve the difficulty encountered in the Klein-Gordon equation stemmed from the fact that it is a second-order differential equation in time. He tried to construct an equation that has linear derivative in time. In order to keep the relativistical covariance, the Hamiltonian should contain only linear spatial derivative as well. Furthermore, the Hamiltonian should also contain the rest mass energy, so that

$$H = c\left(\alpha_1 p_x + \alpha_2 p_y + \alpha_3 p_z\right) + \alpha_0 \mu c^2, \tag{12.50}$$

where $(\alpha_0, \alpha_1, \alpha_2, \alpha_3)$ have to be determined. The Dirac equation reads

$$i\hbar \partial_t \Psi(\mathbf{r},t) = H\Psi(\mathbf{r},t), \tag{12.51}$$

or more explicitly

$$i\hbar \partial_t \Psi(\mathbf{r},t) = \left[c\left(\alpha_1 p_x + \alpha_2 p_y + \alpha_3 p_z\right) + \alpha_0 \mu c^2\right] \Psi(\mathbf{r},t). \tag{12.52}$$

In the discussion below, we will use subscript μ or ν to denote the four components of a vector or matrix, hence subscript μ or ν runs from 0 to 3. We will use subscript i or j to denote the three spatial components of a vector or matrix, hence subscript i or j runs from 1 to 3.

To ensure the Dirac equation relativistically covariant, $\alpha = (\alpha_0, \alpha_1, \alpha_2, \alpha_3)$ should satisfy the following properties:

1. α are time and coordinate independent.
2. H is hermitian, so is α, i.e.

$$\alpha_\mu^\dagger = \alpha_\mu. \tag{12.53}$$

3. The wave function should be the solution of the Klein-Gordon equation so that the eigenenergy and the momentum satisfy Eq. (12.12). Hence we should obtain the Klein-Gordon equation from the equation

$$-\hbar^2 \partial_t^2 \Psi = H^2 \Psi = \left(p^2 c^2 + \mu^2 c^4\right)\Psi, \tag{12.54}$$

which requires

$$c^2\left(\alpha_1 p_x + \alpha_2 p_y + \alpha_3 p_z + \alpha_0 \mu c\right)^2 = p^2 c^2 + \mu^2 c^4. \tag{12.55}$$

This equation is satisfied if and only if

$$\alpha_\mu \alpha_\nu + \alpha_\nu \alpha_\mu = 2\delta_{\mu,\nu}. \tag{12.56}$$

Thus different α_μ anticommute with each other. Since scalars always commute with each other, α_μ cannot not scalars. They must be matrices.

If the dimensions of these matrices are equal to, one can construct three matrices that anticommute with each other. These matrices are just the Pauli matrices. But we cannot find the fourth matrix that anticommutes with the Pauli matrices. Thus the dimension of α_μ cannot be 2.

It is also simple to show that the dimension of α cannot be odd. Otherwise, we will have, for example

$$\alpha_1\alpha_2 = -\alpha_2\alpha_1, \tag{12.57}$$

and

$$\det(\alpha_1\alpha_2) = (-)^N\det(\alpha_2\alpha_1) = (-)^N\det(\alpha_1\alpha_2), \tag{12.58}$$

where N is the dimension of α. The above equation holds only when N is even.

The minimal N that works is 4, which is also what Dirac used. In this case, Ψ is a four-dimensional vector.

These four matrices are not uniquely defined. One popular choice is

$$\alpha_0 = \begin{pmatrix} I & 0 \\ 0 & -I \end{pmatrix}, \quad \alpha_i = \begin{pmatrix} 0 & \sigma_i \\ \sigma_i & 0 \end{pmatrix} \tag{12.59}$$

where $(\sigma_1,\sigma_2,\sigma_3)$ are the Pauli matrices.

In the Dirac equation, Ψ is a four-dimensional vector. It describes the spin states of fermions and is often called a spinor.

12.3.2 CHARGE CONSERVATION

Multiply the Dirac equation (12.52) from left by $\Psi^\dagger$, we obtain

$$i\hbar\Psi^\dagger\partial_t\Psi = -i\hbar c\Psi^\dagger\alpha\cdot\nabla\Psi + \alpha_0\mu c^2\Psi^\dagger\Psi. \tag{12.60}$$

where

$$\alpha = (\alpha_1,\alpha_2,\alpha_3). \tag{12.61}$$

Taking the hermitian conjugate, Eq. (12.60) becomes

$$-i\hbar\left(\partial_t\Psi^\dagger\right)\Psi = i\hbar c\left(\nabla\Psi^\dagger\right)\cdot\alpha\Psi + \alpha_0\mu c^2\Psi^\dagger\Psi. \tag{12.62}$$

Subtracting Eq. (12.60) by Eq. (12.62) then gives

$$\begin{aligned} i\hbar\partial_t\left(\Psi^\dagger\Psi\right) &= -i\hbar c\Psi^\dagger\alpha\cdot\nabla\Psi - i\hbar c\left(\nabla\Psi^\dagger\right)\cdot\alpha\Psi \\ &= -i\hbar c\nabla\cdot\Psi^\dagger\alpha\Psi. \end{aligned}$$

This leads to the current conservation equation

$$\partial_t\left(\Psi^\dagger\Psi\right) + c\nabla\cdot\Psi^\dagger\alpha\Psi = 0, \tag{12.63}$$

which suggests us to define the charge density and current operators as

$$\rho(r,t) = \Psi^\dagger\Psi, \tag{12.64}$$
$$\mathbf{j}(r,t) = c\Psi^\dagger\alpha\Psi. \tag{12.65}$$

The density operator is the square of the absolute magnitude of the spinor wave function, which is positive definite.

The current does not contain any derivatives, unlike the nonrelativistic case. Nevertheless, $c\alpha$ can be regarded as the velocity operator of relativistic electrons

$$\frac{\partial\mathbf{r}}{\partial t} = \frac{1}{i\hbar}[\mathbf{r},H] = c\alpha. \tag{12.66}$$

12.3.3 COVARIANT FORM AND GAMMA MATRICES

The Dirac equation can be written in a covariant form using gamma matrices. To do this, let us multiply both sides of Eq. (12.52) by α_0 and express the Dirac equation as

$$\left(\frac{i\hbar\alpha_0}{c}\partial_t - \alpha_0\alpha_1 p_x - \alpha_0\alpha_2 p_y - \alpha_0\alpha_3 p_z - \mu c\right)\Psi(\mathbf{r},t) = 0. \qquad (12.67)$$

Now we introduce the four gamma matrices $\gamma = (\gamma_0, \gamma_1, \gamma_2, \gamma_3)$

$$\gamma_0 = \alpha_0 = \begin{pmatrix} I & 0 \\ 0 & -I \end{pmatrix}, \qquad (12.68)$$

$$\gamma_i = \alpha_0\alpha_i = \begin{pmatrix} I & 0 \\ 0 & -I \end{pmatrix}\begin{pmatrix} 0 & \sigma_i \\ \sigma_i & 0 \end{pmatrix} = \begin{pmatrix} 0 & \sigma_i \\ -\sigma_i & 0 \end{pmatrix}. \qquad (12.69)$$

The Dirac equation can now be shortened as

$$(\gamma \cdot p - \mu c)\Psi(r,t) = 0, \qquad (12.70)$$

where p is a vector of four momenta defined by

$$p = (p_0, \mathbf{p}), \qquad p_0 = \frac{i\hbar}{c}\partial_t.$$

$\gamma \cdot p$ is a scalar defined by

$$\gamma \cdot p \equiv \gamma_0 p_0 - \gamma_1 p_x - \gamma_2 p_y - \gamma_3 p_x. \qquad (12.71)$$

The gamma matrices have the following properties:

1. γ_0 is hermitian and γ_i is antihermitian

$$\gamma_0^\dagger = \gamma_0, \qquad (12.72)$$

$$\gamma_i^\dagger = (\alpha_0\alpha_i)^\dagger = \alpha_i\alpha_0 = -\gamma_i, \qquad (12.73)$$

and

$$\gamma_0^2 = I, \qquad (12.74)$$

$$\gamma_i^2 = -I. \qquad (12.75)$$

2. Different γ_μ anticommute with each other

$$\gamma_\mu\gamma_\nu + \gamma_\nu\gamma_\mu = 0 \quad (\mu \neq \nu). \qquad (12.76)$$

The above properties can be summarized by one equation

$$\gamma_\mu\gamma_\nu + \gamma_\nu\gamma_\mu = (4\delta_{\mu,0} - 2)\delta_{\mu,\nu}. \qquad (12.77)$$

12.3.4 COUPLED WITH ELECTROMAGNETIC FIELDS

It is straightforward to add electromagnetic interactions to the Dirac equation. Similar as in classical mechanics, we simply substitute E by $E - e\varphi$ and $\mathbf{p}$ by $\mathbf{p} - e\mathbf{A}$, where φ is the scalar potential and $\mathbf{A}$ is the vector potential. In covariant form, it is

$$p_\mu \to p_\mu - eA_\mu, \tag{12.78}$$

where

$$A_\mu = (\varphi, \mathbf{A}). \tag{12.79}$$

The Dirac equation now becomes

$$[\gamma \cdot (p - eA) - \mu c]\Psi(r,t) = 0. \tag{12.80}$$

It can be also written as

$$i\hbar \frac{\partial}{\partial t}\Psi(r,t) = H\Psi(r,t), \tag{12.81}$$

where the Hamiltonian is given by

$$H = c\alpha \cdot (\mathbf{p} - e\mathbf{A}) + e\varphi + \alpha_0 \mu c^2.$$

12.3.5 FREE-PARTICLE SOLUTIONS

We consider the solution of the Dirac equation. For a free particle, the plane wave solution can be expressed as

$$\Psi(r,t) = \psi \exp\left(\frac{i\mathbf{p} \cdot \mathbf{r} - iEt}{\hbar}\right), \tag{12.82}$$

where ψ is a four-dimensional vector independent of space and time. We define the spinor ψ as

$$\psi = \begin{pmatrix} \theta \\ \phi \end{pmatrix} \tag{12.83}$$

with both θ and ϕ two-dimensional spinors. In terms of this notation, the stationary Dirac equation becomes

$$\begin{pmatrix} \mu c^2 & c\sigma \cdot \mathbf{p} \\ c\sigma \cdot \mathbf{p} & -\mu c^2 \end{pmatrix} \begin{pmatrix} \theta \\ \phi \end{pmatrix} = E \begin{pmatrix} \theta \\ \phi \end{pmatrix}, \tag{12.84}$$

where

$$\sigma \cdot \mathbf{p} = \begin{pmatrix} p_3 & p_1 - ip_2 \\ p_1 + ip_2 & -p_3 \end{pmatrix}. \tag{12.85}$$

Eq. (12.84) can be rewritten as

$$c\sigma \cdot \mathbf{p}\phi = (E - \mu c^2)\theta, \tag{12.86}$$
$$c\sigma \cdot \mathbf{p}\theta = (E + \mu c^2)\phi. \tag{12.87}$$

Solving these equations, we find that

$$\phi = \frac{c\sigma \cdot \mathbf{p}}{E + \mu c^2}\theta, \qquad (12.88)$$

and

$$\left[E^2 - \mu^2 c^4 - c^2 (\sigma \cdot \mathbf{p})^2\right]\theta = 0. \qquad (12.89)$$

Thus the eigenenergie is determined by the equation

$$E^2 = \mu^2 c^4 - c^2 (\sigma \cdot \mathbf{p})^2 = \mu^2 c^4 + c^2 \mathbf{p}^2. \qquad (12.90)$$

Here the identity

$$(\sigma \cdot \mathbf{p})^2 = \sum_{ij} \sigma_i \sigma_j p_i p_j = \mathbf{p}^2 \qquad (12.91)$$

is used. Eq. (12.90) is nothing but the Einstein mass-energy equation.

The energy E can be either positive or negative. Let us consider these two cases separately.

1. $E = \sqrt{\mu^2 c^4 + c^2 \mathbf{p}^2} > 0$

   In this case

$$\phi = \frac{c\sigma \cdot \mathbf{p}}{E + \mu c^2}\theta. \qquad (12.92)$$

Since $E > \mu c^2$, the denominator is always larger than the numerator. The θ term is the large component of the wave function and ϕ the small one. Therefore, the wave function is

$$\psi = N \begin{pmatrix} (E + \mu c^2)\theta \\ c\sigma \cdot \mathbf{p}\theta \end{pmatrix}. \qquad (12.93)$$

N is the normalization constant, which is determined by

$$1 = \Psi^\dagger \Psi = N^2 \left[(E + \mu c^2)^2 \theta^\dagger \theta + c^2 \theta^\dagger (\sigma \cdot \mathbf{p})^2 \theta\right]. \qquad (12.94)$$

Assuming $\theta^\dagger \theta = 1$, we find that

$$N = \frac{1}{\sqrt{2E(E + \mu c^2)}}, \qquad (12.95)$$

and

$$\Psi(r,t) = N \begin{pmatrix} (E + \mu c^2)\theta \\ c\sigma \cdot \mathbf{p}\theta \end{pmatrix} \exp\left(\frac{i\mathbf{p}\cdot\mathbf{r} - iEt}{\hbar}\right). \qquad (12.96)$$

For this solution, the probability density is

$$\rho(r,t) = \Psi^\dagger \Psi = 1. \qquad (12.97)$$

The corresponding current density is

$$\mathbf{j}(r,t) = c\Psi^\dagger \alpha \Psi = \frac{c^2 \mathbf{p}}{E}. \qquad (12.98)$$

Thus the current is given by the charge times the relativistic velocity.

2. $E < 0$

In this case,

$$\theta = -\frac{c\boldsymbol{\sigma} \cdot \mathbf{p}}{|E| + \mu c^2}\phi. \tag{12.99}$$

ϕ becomes the large component of the wave function and θ becomes the small component. The wave function is

$$\psi = N \begin{pmatrix} -c\boldsymbol{\sigma} \cdot \mathbf{p}\phi \\ (|E| + \mu c^2)\phi \end{pmatrix}. \tag{12.100}$$

Assuming $\phi^\dagger \phi = 1$, the normalization constant N is still given by

$$N = \frac{1}{\sqrt{2|E|(|E| + \mu c^2)}}$$

The full eigenfunction is

$$\Psi(r,t) = N \begin{pmatrix} -c\boldsymbol{\sigma} \cdot \mathbf{p}\phi \\ (|E| + \mu c^2)\phi \end{pmatrix} \exp\left(\frac{i\mathbf{p} \cdot \mathbf{r} + iE_p t}{\hbar}\right). \tag{12.101}$$

For this solution, the probability and current densities are

$$\rho(r,t) = 1, \tag{12.102}$$

$$\mathbf{j}(r,t) = -\frac{c^2 \mathbf{p}}{|E|}. \tag{12.103}$$

12.3.6 SPIN: ROTATIONAL SYMMETRY

A particle that satisfies the Dirac equation possesses an intrinsic angular momentum, called spin. This is an important property predicted by Dirac.

In nonrelativistic quantum mechanics, the orbital angular momentum $\mathbf{L} = \mathbf{r} \times \mathbf{p}$ commutes with the Hamiltonian with rotational symmetry, hence is conserved. This relies on the fact that $\mathbf{L}$ commutes with the kinetic energy which is proportional to $\mathbf{p}^2$.

In the Dirac equation, the Hamiltonian depends on the momentum linearly. The orbital angular momentum does not commute with the Hamiltonian any more. Instead, for the Hamiltonian defined by Eq. (12.50), the commutation between the orbital angular moment and the Hamiltonian is

$$\begin{aligned} [L_i, H] &= [\varepsilon_{ijk} r_j p_k, c\alpha_l p_l] = c\varepsilon_{ijk}\alpha_l [r_j, p_l] p_k \\ &= i\hbar c\varepsilon_{ijk}\alpha_j p_k = i\hbar c (\boldsymbol{\alpha} \times \mathbf{p})_i. \end{aligned} \tag{12.104}$$

Here the summation for repeated subscript indices is implicitly assumed. The above equation indicates that the orbital angular momentum is not conserved by the Dirac equation. It implies that the angular momentum should contain an extra term if the Hamiltonian is invariant under rotation. Let us assume the angular momentum to be

$$\mathbf{J} = \mathbf{L} + \mathbf{S}, \tag{12.105}$$

where S is an intrinsic angular momentum independent on r and p. The commutator of J_i with H is given by

$$
\begin{aligned}
[J_i, H] &= [L_i, H] + [S_i, H] \\
&= i\hbar c \varepsilon_{ijk} \alpha_j p_k + [S_i, c\alpha_j p_j + \alpha_0 \mu c^2] \\
&= i\hbar c \varepsilon_{ijk} \alpha_j p_k + c[S_i, \alpha_j] p_j + \mu c^2 [S_i, \alpha_0]. \quad (12.106)
\end{aligned}
$$

By requiring J to be conserved operator so that it commutes with the Hamiltonian,

$$
[J, H] = 0, \quad (12.107)
$$

we find the equations that S should satisfy

$$
[S_i, \alpha_0] = [S_i, \alpha_i] = 0, \quad (12.108)
$$
$$
[S_i, \alpha_j] = i\hbar \varepsilon_{ijk} \alpha_k. \quad (12.109)
$$

It turns out that the solution is very simply

$$
S = \frac{\hbar}{2} \begin{pmatrix} \sigma & 0 \\ 0 & \sigma \end{pmatrix}. \quad (12.110)
$$

This is nothing but the spin operator of $S = 1/2$.

It can be also shown that $J = L + S$, instead of L itself, is the generator of an infinitesimal rotation in space for the Dirac equation.

12.3.7 ANTIPARTICLES: CHARGE CONJUGATE

The eigenenergy of the Dirac Hamiltonian can be either positive or negative. The negative energy is related to the eigenstates for the antiparticles. Dirac predicted the existence of antiparticles before they were discovered. In fact, the Dirac theory was accepted only after the discovery of the positron.

Electron is a particle of negative charge and positive mass. Positron is a particle with positive charge and the same mass. Positron is the antiparticle of electron. The Dirac equations for these two particles are

$$
[\gamma \cdot (p - eA) - \mu c] \Psi_e(r, t) = 0, \quad (12.111)
$$
$$
[\gamma \cdot (p + eA) - \mu c] \Psi_p(r, t) = 0, \quad (12.112)
$$

where Ψ_e and Ψ_p are the wave functions of electron and positron, respectively. These two equations are not independent. They are related by the transformation of charge conjugation.

Let us first take the complex conjugate for Eq. (12.111). The complex conjugate gives $p^* = -p$ and $A^* = A$ for all four components:

$$
[-\gamma^* \cdot (p + eA) - \mu c] \Psi_e^* = 0. \quad (12.113)
$$

For the four γ matrices, γ_1, γ_3, and γ_4 are real, but γ_2 is purely imaginary

$$\gamma_2^* = -\gamma_2, \tag{12.114}$$
$$\gamma_\mu^* = \gamma_\mu \quad (\mu = 1,3,4). \tag{12.115}$$

These γ-matrices have the following property

$$\gamma_2\gamma_\mu^*\gamma_2 = \gamma_\mu. \tag{12.116}$$

This is because $\gamma_2^2 = -I$ and

$$\gamma_2\gamma_2^*\gamma_2 = -\gamma_2\gamma_2\gamma_2 = \gamma_2, \tag{12.117}$$
$$\gamma_2\gamma_\mu^*\gamma_2 = \gamma_2\gamma_\mu\gamma_2 = -\gamma_2\gamma_2\gamma_\mu = \gamma_\mu, \quad (\mu \neq 2). \tag{12.118}$$

Multiplying γ_2 from left to Eq. (12.113) yields

$$[\gamma \cdot (p + eA) - \mu c]\gamma_2\Psi_e^* = 0, \tag{12.119}$$

which is just the equation for the positron if we set

$$\Psi_p = -i\gamma_2\Psi_e^*. \tag{12.120}$$

This is just the transformation of the wave function under the charge conjugation. $-i\gamma_2$ is the charge conjugate operator. The phase factor of $-i$ is added for convenience because $-i\gamma_2$ is a real matrix.

If

$$\Psi_e = \begin{pmatrix} \Psi_1 & \Psi_2 & \Psi_3 & \Psi_4 \end{pmatrix}^T \tag{12.121}$$

is the wave function of electron, then the corresponding positron wave function is

$$\Psi_p = \begin{pmatrix} 0 & 0 & 0 & -1 \\ 0 & 0 & 1 & 0 \\ 0 & 1 & 0 & 0 \\ -1 & 0 & 0 & 0 \end{pmatrix} \begin{pmatrix} \Psi_1^* \\ \Psi_2^* \\ \Psi_3^* \\ \Psi_4^* \end{pmatrix} = \begin{pmatrix} -\Psi_4^* \\ \Psi_3^* \\ \Psi_2^* \\ -\Psi_1^* \end{pmatrix}. \tag{12.122}$$

12.3.8 NEGATIVE ENERGY: DIRAC SEA

Suppose Ψ_e to be an eigenstate of H with an eigenenergy E, i.e.

$$(c\alpha \cdot \mathbf{p} + \alpha_0\mu c^2)\Psi_e = E\Psi_e, \tag{12.123}$$

now let us consider the energy of the corresponding positron state Ψ_p defined by Eq. (12.120). Take the complex conjugate for this equation, we obtain

$$\gamma_2\left(-c\alpha^* \cdot \mathbf{p} + \alpha_0\mu c^2\right)\Psi_e^* = E\gamma_2\Psi_e^*, \tag{12.124}$$

It can be also written as

$$\gamma_2\left(-c\alpha^* \cdot \mathbf{p} + \alpha_0\mu c^2\right)\gamma_2\Psi_p = -E\Psi_p, \tag{12.125}$$

From the definition, it is simple to show that

$$\gamma_2 \alpha_0 \gamma_2 \ = \ \alpha_0, \tag{12.126}$$

$$\gamma_2 \alpha_i^* \gamma_2 \ = \ \gamma_2 \alpha_1 \gamma_2 = -\alpha_i. \tag{12.127}$$

Thus Eq. (12.125) is

$$(c\alpha \cdot \mathbf{p} + \alpha_0 \mu c^2) \Psi_p = -E\Psi_p. \tag{12.128}$$

Hence Ψ_p is also an eigenstate of H, but with an energy $-E$.

Thus the eigenstate of positron, Ψ_p, is a negative energy state of electron, Ψ_e. In the ground state, the positron states are completely filled. These filled states are called the Dirac Fermi sea.

12.4   NONRELATIVISTIC LIMIT OF THE DIRAC EQUATION

Now let us consider the nonrelativistic limit of the stationary Dirac equation

$$[c\alpha \cdot (\mathbf{p} - e\mathbf{A}) + V(r) + \alpha_0 \mu c^2] \Psi = E\Psi \tag{12.129}$$

when the momentum of a massive particle is small in comparison with μc, i.e. the limit (assuming $E > 0$)

$$E' = E - \mu c^2 \ll \mu c^2. \tag{12.130}$$

$V(r)$ is the potential energy.

Using the two spinors representation of the wave function

$$\Psi(\mathbf{r}, t) = \begin{pmatrix} \theta(\mathbf{r}, t) \\ \phi(\mathbf{r}, t) \end{pmatrix}, \tag{12.131}$$

we can write the Dirac equation as

$$\begin{pmatrix} V + \mu c^2 & c\sigma \cdot (\mathbf{p} - e\mathbf{A}) \\ c\sigma \cdot (\mathbf{p} - e\mathbf{A}) & V - \mu c^2 \end{pmatrix} \begin{pmatrix} \theta \\ \phi \end{pmatrix} = E \begin{pmatrix} \theta \\ \phi \end{pmatrix}, \tag{12.132}$$

or

$$(V - E') \theta + c\sigma \cdot (\mathbf{p} - e\mathbf{A}) \phi \ = \ 0, \tag{12.133}$$

$$c\sigma \cdot (\mathbf{p} - e\mathbf{A}) \theta + (V - E' - 2\mu c^2) \phi \ = \ 0. \tag{12.134}$$

From Eq. (12.134), we find that

$$\phi = \frac{c}{2\mu c^2 - V + E'} \sigma \cdot (\mathbf{p} - e\mathbf{A}) \theta. \tag{12.135}$$

In the nonrelativistic limit, ϕ is a small component in comparison with θ. Substituting it into Eq. (12.133) gives

$$\left[V + c^2 \sigma \cdot (\mathbf{p} - e\mathbf{A}) \frac{1}{2\mu c^2 - V + E'} \sigma \cdot (\mathbf{p} - e\mathbf{A}) \right] \theta = E'\theta. \tag{12.136}$$

12.4.1   FIRST ORDER APPROXIMATION

In Eq. (12.136), if we ignore V and E' in the denominator, it becomes

$$\left[V + \frac{1}{2\mu} \sigma \cdot (\mathbf{p} - e\mathbf{A}) \, \sigma \cdot (\mathbf{p} - e\mathbf{A}) \right] \theta = E' \theta. \tag{12.137}$$

The second term on the left can be simplified as

$$\sigma \cdot (\mathbf{p} - e\mathbf{A}) \, \sigma \cdot (\mathbf{p} - e\mathbf{A}) = (\mathbf{p} - e\mathbf{A})^2 - e\hbar \sigma \cdot \mathbf{B} \tag{12.138}$$

where

$$\mathbf{B} = \nabla \times \mathbf{A} \tag{12.139}$$

is the magnetic field.

Now the Dirac equation becomes

$$\left[\frac{1}{2\mu} (\mathbf{p} - e\mathbf{A})^2 + V - \mu_B \sigma \cdot \mathbf{B} \right] \theta = E' \theta. \tag{12.140}$$

We thus obtain the leading order approximation of the Hamiltonian in the nonrelativistic limit

$$H^{(1)} = \frac{1}{2\mu} (\mathbf{p} - e\mathbf{A})^2 + V - \mu_B \sigma \cdot \mathbf{B}, \tag{12.141}$$

where

$$\mu_B = \frac{e\hbar}{2\mu} \tag{12.142}$$

is the intrinsic magnetic moment of the Dirac particle. The last term is the Pauli interaction between the particle spin and the external magnetic field. For electrons, the measured magnetic moment is

$$\mu_{\text{electron}} = 1.00116 \mu_B, \tag{12.143}$$

close to what predicted by Dirac.

12.4.2   SECOND ORDER APPROXIMATION

The second order approximation is mainly the contribution from the kinetic energy and the scalar potential. We ignore the vector potential for saving effort in the derivation. The Dirac equation now becomes

$$\left(V - E' + c^2 \sigma \cdot \mathbf{p} \frac{1}{2\mu c^2 - V + E'} \sigma \cdot \mathbf{p} \right) \theta = 0. \tag{12.144}$$

The second order approximation is obtained by expanding the denominator on the left to the order of $E' - V$,

$$\frac{1}{2\mu c^2 - V + E'} \approx \frac{1}{2\mu c^2} \left(1 - \frac{E' - V}{2\mu c^2} \right). \tag{12.145}$$

Put it into the above Dirac equation, we have

$$\left(V+\frac{1}{2\mu}\mathbf{p}^2+\frac{1}{4\mu^2c^2}\sigma\cdot\mathbf{p}V\sigma\cdot\mathbf{p}\right)\theta=\left(1+\frac{\mathbf{p}^2}{4\mu^2c^2}\right)E'\theta. \tag{12.146}$$

The third term on the left can be simplified using the formula

$$\sigma\cdot\mathbf{p}V=V\sigma\cdot\mathbf{p}-i\hbar\sigma\cdot(\nabla V), \tag{12.147}$$

and

$$\begin{aligned}\sigma\cdot\mathbf{p}V\sigma\cdot\mathbf{p}&=V(\sigma\cdot\mathbf{p})^2-i\hbar\sigma\cdot(\nabla V)(\sigma\cdot\mathbf{p})\\&=V\mathbf{p}^2-i\hbar(\nabla V)\cdot\mathbf{p}+\hbar\sigma\cdot(\nabla V)\times\mathbf{p}.\end{aligned} \tag{12.148}$$

Substituting this expression into Eq. (12.146), we obtain the stationary Dirac equation up to the second order approximation

$$\begin{aligned}&\frac{1}{2\mu}\mathbf{p}^2\theta+\frac{1}{4\mu^2c^2}\left[\hbar\sigma\cdot(\nabla V)\times\mathbf{p}-i\hbar(\nabla V)\cdot\mathbf{p}\right]\theta\\&=(E'-V)\left(1+\frac{\mathbf{p}^2}{4\mu^2c^2}\right)\theta.\end{aligned} \tag{12.149}$$

12.4.3 NORMALIZATION

In relativistic physics, the normalization of the wave function is

$$\Psi^\dagger\Psi=\theta^\dagger\theta+\phi^\dagger\phi=1. \tag{12.150}$$

Taking the leading order approximation for the small component in Eq. (12.135), we find that

$$\phi=\frac{1}{2\mu c}\sigma\cdot\mathbf{p}\theta. \tag{12.151}$$

Thus the normalization condition for θ is given by

$$\theta^\dagger\left[1+\frac{(\sigma\cdot\mathbf{p})^2}{4\mu^2c^2}\right]\theta=\theta^\dagger\left(1+\frac{\mathbf{p}^2}{4\mu^2c^2}\right)\theta=1. \tag{12.152}$$

This implies that the 2-dimensional spinor θ' defined by

$$\theta'=\left(1+\frac{\mathbf{p}^2}{4\mu^2c^2}\right)^{1/2}\theta\approx\left(1+\frac{\mathbf{p}^2}{8\mu^2c^2}\right)\theta$$

is normalized to 1

$$\theta'^\dagger\theta'=1. \tag{12.153}$$

12.4.4 EFFECTIVE HAMILTONIAN

Keeping the terms up to the order μ^{-3}, the Dirac equation for θ' can be written as

$$\frac{1}{4\mu^2 c^2} \left[\hbar\sigma \cdot (\nabla V) \times \mathbf{p} - i\hbar\,(\nabla V) \cdot \mathbf{p} \right] \theta'$$

$$= \left[(E' - V) \left(1 + \frac{\mathbf{p}^2}{8\mu^2 c^2} \right) - \frac{1}{2\mu} \mathbf{p}^2 \left(1 - \frac{\mathbf{p}^2}{8\mu^2 c^2} \right) \right] \theta' \qquad (12.154)$$

It can be expressed as

$$H^{(2)} \theta' = E' \theta', \qquad (12.155)$$

where $H^{(2)}$ is the Hamiltonian for θ'. Up to the terms of the order μ^{-3}, it is given by

$$
\begin{aligned}
H^{(2)} &= \frac{1}{2\mu} \mathbf{p}^2 \left(1 - \frac{\mathbf{p}^2}{4\mu^2 c^2} \right) + V' \\
&+ \frac{\hbar}{4\mu^2 c^2} \sigma \cdot (\nabla V) \times \mathbf{p} - \frac{i\hbar}{4\mu^2 c^2} (\nabla V) \cdot \mathbf{p}, \qquad (12.156)
\end{aligned}
$$

where

$$
\begin{aligned}
V' &= \left(1 - \frac{\mathbf{p}^2}{8\mu^2 c^2} \right) V \left(1 + \frac{\mathbf{p}^2}{8\mu^2 c^2} \right) \\
&\approx V + \frac{\hbar^2}{8\mu^2 c^2} \nabla^2 V + \frac{i\hbar}{4\mu^2 c^2} (\nabla V) \cdot \mathbf{p}. \qquad (12.157)
\end{aligned}
$$

We therefore have

$$H^{(2)} = \frac{\mathbf{p}^2}{2\mu} + V - \frac{\mathbf{p}^4}{8\mu^3 c^2} + \frac{\hbar^2}{8\mu^2 c^2} \nabla^2 V + \frac{\hbar}{4\mu^2 c^2} \sigma \cdot (\nabla V) \times \mathbf{p}. \qquad (12.158)$$

In a central potential, $V = V(r)$, it becomes

$$H^{(2)} = \frac{\mathbf{p}^2}{2\mu} + V - \frac{\mathbf{p}^4}{8\mu^3 c^2} + \frac{\hbar^2}{8\mu^2 c^2} \nabla^2 V + \frac{\hbar}{4\mu^2 c^2 r} \frac{dV}{dr} \sigma \cdot \mathbf{L}. \qquad (12.159)$$

It contains three correction terms:

1. The third term is the correction to the kinetic energy, which is derived from the expansion

$$\sqrt{\mu^2 c^4 + \mathbf{p}^2 c^2} \approx \mu c^2 + \frac{\mathbf{p}^2}{2\mu} - \frac{\mathbf{p}^4}{8\mu^3 c^2} + \cdots. \qquad (12.160)$$

2. The fourth term is called the Darwin term. If V is a Coulomb potential, then

$$\nabla^2 \frac{1}{r} = -4\pi\delta(r) \qquad (12.161)$$

is a delta function at the origin. It affects only the s-wave states whose wave functions are finite at the origin, since the wave function of an electron with $l > 0$ vanishes at the origin. It leads to the Fermi contact interaction between an electron and an atomic nucleus, which is responsible for the appearance of isotropic hyperfine coupling.

3. The fifth term is the spin-orbit interaction. It affects only non-s-states with finite orbital angular momenta.

12.5 PROBLEMS

1. Verify that the solution of the Dirac equation is also a solution of the Klein-Gordon equation.
2. Show that the Dirac Hamiltonian commutes with the total angular momentum operator $\mathbf{J} = \mathbf{L} + \mathbf{S}$.
3. An electron is confined to the x-y plane. Solve for the eigenvalue and eigenvector of the Dirac equation for an electron in a constant magnetic field along the z-axis.
4. Consider two relativistic corrections to an electron in a three-dimensional harmonic oscillator

$$V(\mathbf{r}) = \frac{1}{2}\mu\omega^2\mathbf{r}^2.$$

   a. Evaluate the Darwin term.
   b. Evaluate the spin–orbit interaction term.
5. For a free electron, find the common eigenstates of H, $\mathbf{p}$, and $\mathbf{p}\cdot\mathbf{S}$. $\mathbf{S}$ is the spin operator.
6. Ehrenfest's theorem: $\Psi(\mathbf{r},t)$ is a normalized solution of the Dirac equation. Show that the expectation value of an arbitrary spinor operator O satisfies the equation

$$\frac{d}{dt}\langle\Psi|O|\Psi\rangle = \frac{1}{i\hbar}\langle\Psi|[O,H]|\Psi\rangle + \langle\Psi|\partial_t O|\Psi\rangle.$$

This is just a generalization of the Ehrenfest theorem to the relativistic system.

Index

Printed in the United States
by Baker & Taylor Publisher Services